Conference on

Safe Water Environments

Eldoret, Kenya, August 21-23, 1995

Dept of Water and Environmental Studies,
Linköping University, Sweden

Faculty of Health Sciences,
Moi University, Kenya

London School of Hygiene and Tropical Medicine, UK

FOREWORD

Decision-makers at all levels need comprehensive accounts of ongoing research and development to reach informed conclusions. The information often is fragmented and inaccessible. Therefore, there are good reasons to present the current state of knowledge in various fields related to periurban sanitation.

Water and sanitation in the household and its immediate neighbourhood formed the subject of this conference on Safe Water Environments. The word "Safe" points to water´s direct effect on people's health rather than to specific standards of water. We did not choose a priority: like health for all, water for all, absence of pathogens. Experience has shown that improvement in health arises from recognising an intricate web of factors. Ambitious, single-minded approaches rarely lead to quick progress, but most often end in disillusionment.

Amid the many issues these papers discuss, an overriding concern is the relationship between human behaviour and pathogen spread. This arises from the common understanding that improved water and sanitation is not a technical and scientific matter only, but a social process as well.

The first three papers centre around conflicts and cooperation in neighbourhoods where drainage is not provided by the municipality. What can tenants do about flies and waste? Moist conditions favour, among other things, survival of pathogens and breeding of mosquitos. Solid waste and faecal deposits provide food for rats. Stormwater and wastewater drainage play a crucial role in keeping down pathogens, but can be undermined by weak institutional arrangements, ambiguous security of tenure, and uncontrolled growth of settlement areas.

The prospects and problems of a holistic research method are illustrated in the next two papers on the Safe Water Environments research programme in Eldoret, Kenya. One paper depicts the social dimension of how residents create their environmental conditions and how they view the results. The other looks at the interplay between humans and pathogens through the correlation between childhood diarrhoea incidence and measured water quality.

The factors motivating residents´ choice of water are the subject of the next paper. Residents may categorize water as "new" or "used" according to their own criteria, rather than accepting professional standards and treatments. Should there be an investment in more water or in better quality water? Health benefits from improved water quality are not as large as expected. However, adequate quality help prevent water-borne epidemics. Policy decisions, to be effective, must attend to local perceptions.

A specific route of faecal contamination is from dug latrines via groundwater into dug wells. Research using tracers of various kinds confirming this well water contamination route leads to recommendations on the safest distances between wells, groundwater and latrines. There is a disucussion of the degree to which people perceive the scientifically established contamination of their water sources.

Most societies have their periurban slums. Examples from around the world are presented in the next subject area. A study of the development of water and sewerage treatment during the industrialization in the North ascertains the weak link between knowledge of health related issues and action taken to resolve them. In today´s world, a large comparative study of Accra, Sao Paulo and Jakarta shows poor periurban areas suffer from enormous variations in the provision of water and sewerage.

Current ways of involving local communities in water and sanitation are evaluated. While the methods were developed for rural areas they are now being applied in urban settings. As they expand, participatory projects will shape life and institutions in local communities where the formal sector is weak. A study of the law in relation to water and sanitation issues in Kenya makes a strong case for keeping the links between rules and social aspirations. Other papers, however, propose potent legal regulations as part of the improvement strategy.

In the panel discussion, decision-makers at national and municipal levels had the opportunity to discuss their own experience of the conference issues. Are there safe alternatives for disposal of human waste in congested areas? Physical planning is vital but seems to come in only after areas are settled and by then it is more difficult and costly. The unanimous view is that legal action is a last resort.

The key note speech and the summing-up session of the conference pointed at important issues to be studied further: a change in knowledge does not necessarily lead to a change in attitude and practice, projects must remain sustainable after outside support ceases. But how?

The conference was a cooperative effort by the speakers and their institutions, Moi University, Linköping University and Sida. The proceedings were edited by Jan-Olof Drangert, Richard Swiderski and Melvin Woodhouse.

List of Content

OPENING SPEECH

AT THE CONFERENCE ON SAFE WATER ENVIRONMENTS

Hon. Joshua M. Angatia, EGH, M.P., Minister of Health, Kenya

Hon. John Sambu, EGH, M.P., Minister for Environment & Natural Resources,
Mr. Chairman, Distinguished delegates, Ladies and Gentlemen.

I consider it a great honour to be given the opportunity to open this important conference. I am, therefore, pleased to be with you here today.

Let me join the organisers in thanking you for being able to attend this conference, and invite you to feel at home in Eldoret. For those who have come from outside Kenya, may I, on behalf of the people and Government of Kenya, welcome you to our country and invite you to spare some time to see a bit of our country, especially those areas which make this country rich in culture and natural beauty.

During the earth summit in Rio de Janeiro, world leaders, who included our beloved President Hon. Daniel Toroitich Arap Moi, affirmed their commitment to environmental protection. Since then concern for the ozone layer has grown tremendously, not only among all of us gathered here, but also among many citizens of Kenya as well as the world. Sustainable development means "adopting business strategies and activities that meet the needs of the enterprise and its shareholders today while protecting, sustaining and enhancing the human and natural resources that will be needed in the future". Business and enterprise as used here refer to the wide spectrum of human activities and endeavour which have an impact on the environment.

At this juncture, let me emphasis that the Government of Kenya is committed to these ideals on environment, as has been amply emphasised by none other than H.E. The President himself, who has established the Ministry of Environment and Natural Resources to pursue the implementation of the Environmental Agenda. I am pleased to note that my colleague Hon. John Sambu, the Minister in charge is here with us.

Mr. Chairman, Ladies and Gentlemen, you know that water is a crucial component of the environment. Water is part of the environment in its distribution, its quantities and its quality. For the people of Kenya, the Government is committed to providing adequate clean water within reasonable distance of their homes. To date, a good proportion of our people have access to affordable good quality water.

For the remaining section of the community which still continues to suffer unavailability or inadequacy of good quality water, the pledge by the Government to continue to address this deprivation remains unwavering despite the harsh economic realities.

This Government is for the people and by the people of this country. So the environment in which the people live is of paramount importance. It is our desire that the people live in an environment which promotes and protects their good health. Despite the fact that most rural areas of Kenya do not have piped water, electricity, tarmac roads, telephone services, etc., the physical environment is healthier in these rural areas than it is in the urban areas which tend to suffer from all sorts of pollution, congestions and contacts.

In this sense, one can say that the phenomenon of men leaving their families in the rural areas to seek wealth in the towns and returning home in the rural areas to use it is justified. I do not know whether this practice is only in Africa, or whether it exists in other communities, especially the third world countries. As we have all heard, and many of you have witnessed, many cities of the world are on the verge of collapse. The infrastructure is over-extended by the exodus of people to the cities. Many cities can no longer provide adequate jobs, housing, transport systems and even supplies of water. And many cities are choked by mountains of waste on the ground, and, or, by air pollution in the atmosphere.

Ladies and Gentlemen, in spite of differences in availability of resources, varying forms of governments, degrees of industrialisation and stages of development of the countries of the world, deterioration of the environment poses a threat to the entire international community.

Therefore:

– What can we do to enable all the people of the world to have equal access to the basic resources?

– What can we do as educators, scientists, researchers, policy makers and implementors to arrest the slide into unsanitary living conditions for our people?

– How can we address the needs of urban communities? and what about the rural communities?

– What are the appropriate technologies required by the various communities?

Let me narrow my passing comments to a few topics which you may discuss at this conference.

Waste disposal was identified as an integral component of the strategies for sustainable development. Governments were urged to launch programmes both "to demonstrate enhanced waste recycling and to make it operational." Key to the realisation of this goal is the establishment of an appropriate infrastructure for the management of waste. Today we are not confronted with such pestilences as plague which were common in Europe and other heavily settled areas in the 18th and 19th centuries. But we are in constant danger of other equally, if not more, deadly epidemics, such as typhoid and cholera. Allow me to add, that human to human pollution, such as transmission of diseases, has become a great threat to mankind.

In Kenya, do we know the main contents of the waste that is being constantly discharged by industry and from households and other institutions? How much of this could be recycled or economically exploited? What technology is required to transform waste from a dangerous enemy to a source of environmental regeneration and future survival of people? Or how might we co-exist with waste?

I note from your conference agenda that you will be addressing various topics ranging from storm water to pests and to city slums; and from scientific principles and concepts to anthropological practices. As you know, these are the areas where your ground becomes common with ours in the Ministry of Health. Malaria still continues to be a major killer in our country. You may be aware of the high rate of resistance of diseases to commonly used drugs. You also know that water-borne diseases, particularly those affecting children, are a major threat of many of our communities. It is my hope that while addressing the problems of defined urban and periurban communities you will also suggest solutions which benefit members of the wider society; that while you propose solutions that are scientifically and medically sound, such solutions that are likely to gain wide acceptance by our society. And that while you can tell the quality of clean water through laboratory tests, you can also provide the community with effective tools upon which to judge the quality of water.

Mr. Chairman, Distinguished Delegates, Ladies and Gentlemen, the concern for the Environment cannot be overemphasized. The consequences of ignoring the urgency of its needs are too grave to contemplate. No single sector of the society has a monopoly of ideas to viable solutions. Your inputs are most important, and eagerly awaited. Let me assure you that the effective and workable solutions which you will propose will be studied and implemented.

It is now my great pleasure, ladies and gentlemen, to declare this National Conference on Safe Water Environments, officially open.

Thank you for your attention

KEYNOTE ADDRESS

WATER AND SANITATION FOR THE YEAR 2000 AND BEYOND

David J. Bradley
London School of Hygiene and Tropical Medicine

Introduction

It is highly appropriate that this conference should be held in Kenya since the country is notable for having taken seriously, over many years, both the provision of improved domestic water supplies and the promotion of medical and health research. The record of Kenya stands high within Africa and among developing countries more broadly. The project which is being presented particularly at this conference and which has been taking place at Moi University provides a very clear example of the interdisciplinary work needed to understand and solve the problems of water and health. My own institution, the London School of Hygiene and Tropical Medicine, much appreciates having been invited to co-sponsor the event. I also notice that the motto of Moi University is particularly appropriate: "Fountains of knowledge" fits well in all senses of the words. There is also personal pleasure because it is 35 years since I first came to live in East Africa and began to work on water related disease myself.

However the topic is an extremely tall order: Water and sanitation for the year 2000 and beyond. Peter Medawar, a great scientist, gave very good reasons why one couldn't give talks predicting advances in science, because if one could predict them one would have made them! 2000 really is quite close to us today and discoveries take time to filter through into practice, so for some time to come we are going to have to use what is currently on the horizon or is already available. Moreover, in order to look forwards it is sometimes helpful to look backwards as well and to give ourselves some kind of perspective. This paper is rather general because of the varied backgrounds of potential readers but aims to progress through issues of health and its relations to water, but to be cautious about providing advice to people who are working on the ground and providing other papers. It was symbolically very appropriate that when this paper was delivered not one but two Kenya Government Ministers were present: the Health Minister and the Minister for the Environment.

Health, Environment and Development

It is quite clear that matters of health, environment and socio-economic development interact continuously to an extreme degree. One cannot usefully think about one without considering the others, but it is remarkable to have this demonstrated quite so clearly by having two ministers present. One could detail the various interactions but I would particularly like to point out that there are two sets of environmental changes going on. One set comes into the category of environmental health in the conventional sense, that is to say, altering the environment specifically to improve health: things like water supply and sanitation. But equally important are environmental changes taking place in the interests of socio-economic development. They may have direct effects, both good and bad, on health and they also may increase income, hopefully of many people (Fig 1). They may even reduce the income of a few. All those changes in income also have an effect on health, so we are dealing with a very complicated situation, but one that is full of interest. It is very important to think about the developments which are not *primarily* for the sake of health and to ensure that they too contribute to the improvement of health.

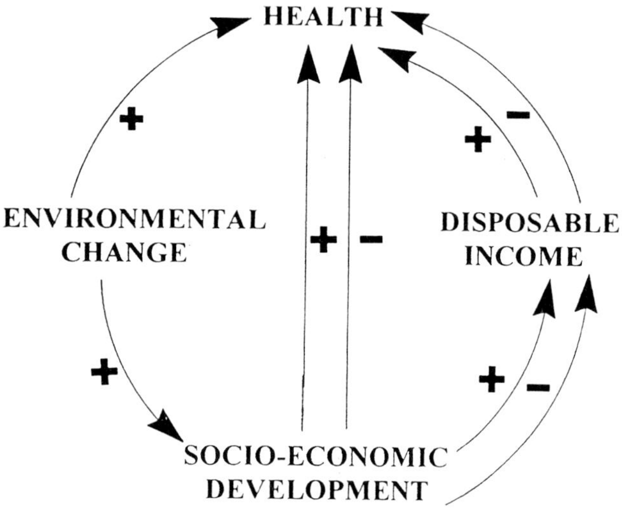

Fig. 1. Effects of environmental changes upon health

If we look at mortality and health in the world as a whole, Fig 2 gives the age distribution of people who are dying. There is a marked contrast between the two circles: in the industrialised countries only 4% of the total deaths are under 14 years of age, whereas 40% in developing countries are from that age group. That is a consequence of two things: the nature of health problems of developing countries and the different age distribution of the two populations. The brunt of those illnesses falls particularly on our children. Also there are proportionately more children, so the population issues are intimately involved with health issues and the differences that are occurring. If we trace back to the causes of death of those children and the reasons for the big differences in mortality, many of them can be traced back to two linked health problems. One is malnutrition, addressed by others, and the other is infectious disease. Infections in turn are the result of two main things: that group of phenomena that we call poverty, which covers a whole range of things, and secondly warm climates. There are some diseases which depend on a warm climate for parts of their life cycles, such as malaria and schistosomiasis, but in the case of both categories there are close links between ill health and water and equally between improvements in health and various changes in the way we interact with water.

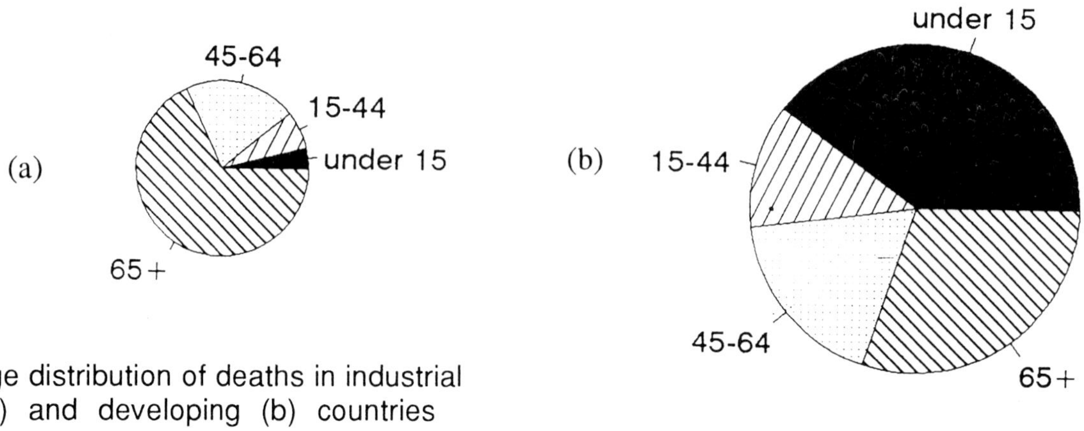

Fig. 2. Age distribution of deaths in industrial (a) and developing (b) countries

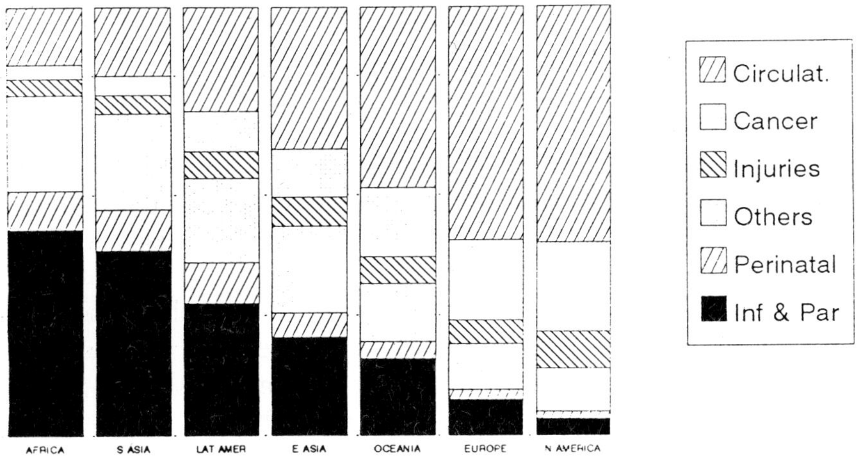

Fig. 3. The relative importance of different causes of death in the world by continent

Fig. 3 emphasises in quantitative terms the importance of infectious and parasitic diseases in Africa, where they preponderate to a greater degree than elsewhere in the world and where they present us with a bleak challenge but one which it is possible to overcome. While we do not know how to tackle some diseases, in the case of the infections, and particularly the water related ones, we at least know what we are trying to do.

Fig 4 illustrates the morbidity problems of developing countries with data from Guatemala, from the most comprehensive study of its kind, following the growth of children in a village. Fig. 4 plots the growth of one of those children. The interrupted line shows the growth trajectory that would have been followed had that child been well off, living in the capital city. The solid line shows the actual growth together with all the various things that happened

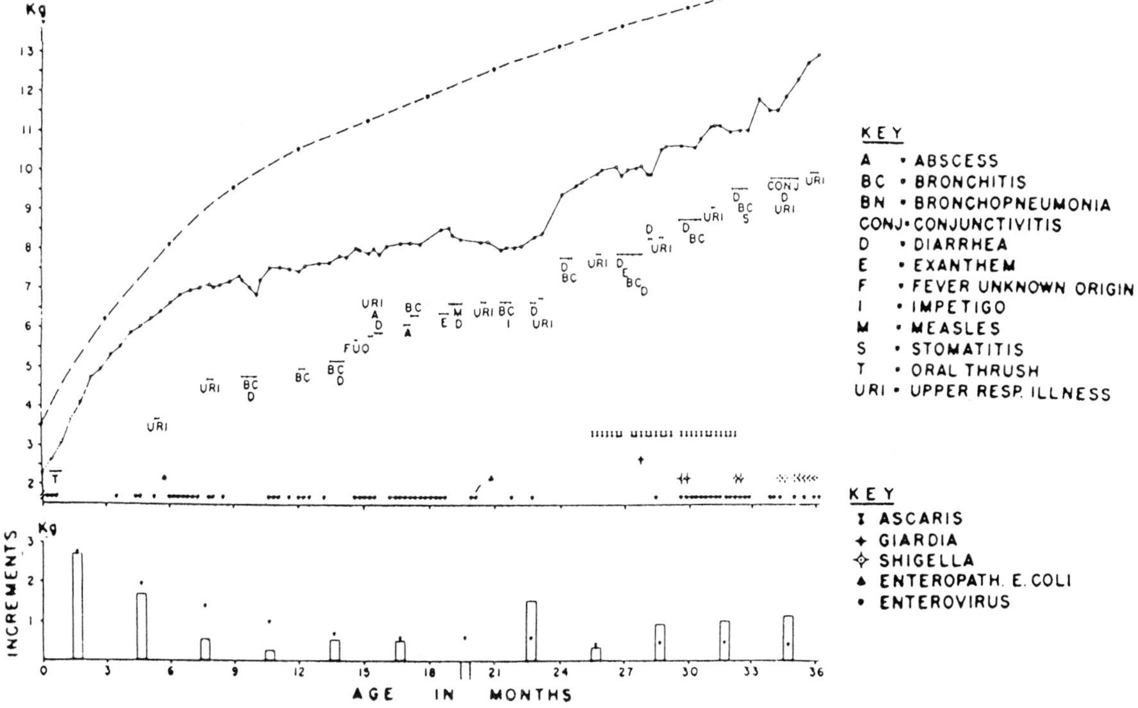

Fig. 4. Growth of a child in a village in Guatemala *(L.J. Mata, The children of Santa Maria Cauque)*

to the child: attacks of diarrhoea, fevers, a whole range of infections and diverse parasites. The consequence was that that child illustrated scarcely increased in weight between the ages of 6 months and 3 years. Since for a child growth=health we have a very clear example of the consequences of the many infections, most of which are related either directly or indirectly to water, affecting the growth and health of children. The situation in Africa is even less satisfactory.

Water Use in East Africa

30 years ago I was fortunate enough to be first involved in domestic water supply studies within East Africa together with Professor Gilbert White and Anne White, geographers with a deep knowledge of water, decision making and environmental risk, in a multidisciplinary study on 30 sites throughout East Africa: 10 each in Kenya, Uganda and Tanzania, covering a whole range of different levels of living, urban and rural, to find and determine for the first time how much water people actually used domestically, what were the characteristics of it, and how decisions about water use were made. The fieldwork was done, as was the case for some of the studies reported at this conference, with medical students and geography students during their long vacation and quite a number of those have gone on to occupy senior posts within East Africa, which may perhaps encourage those doing the fieldwork in connection with the Moi University study!

Some of the findings are briefly presented in case some of you are not familiar with them. The first is the dramatic range in water use by different groups of people, from a low of 2.5 litres per capita per day up to over 600 litres per capita per day, the latter in a rich area of Dar es Salaam, and the former in a mountainous area of Uganda. Within the whole

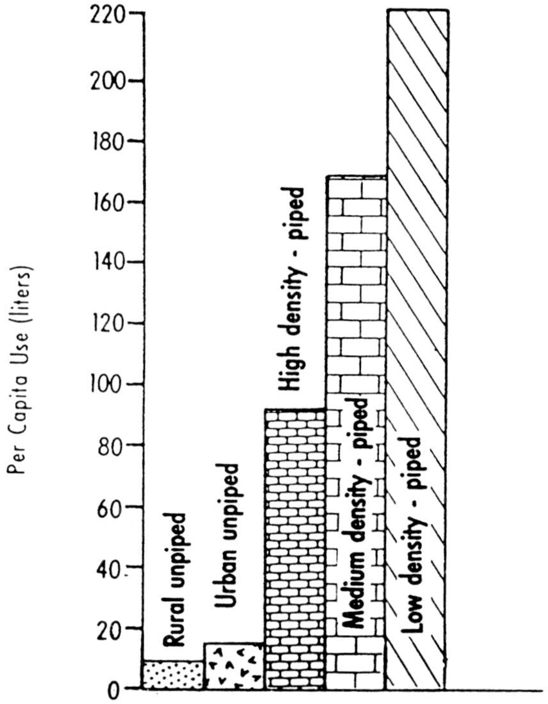

Fig. 5. Water use in East Africa by site

of East Africa there is great variation, obviously much lower in people who have to carry their water than among those who have water piped into their compound (Fig. 5). So there is enormous diversity of availability of water. Rather interestingly, and contrary to what had been believed until then, when the distance to water is from a very small distance from the house up to nearly a kilometre, there is rather little variation in the amount of water that people carry, though use falls off for greater distances. But the per capita use if water is being carried rarely exceeds 10 litres per head per day which is substantially under what is desirable for adequate health. If people move from rural to urban areas their water use rises somewhat.

Perhaps the most worrying conclusion, and also the one that gives us most hope for the future, is that poor people, particularly in urban areas, often spend a very large proportion of their incomes on buying water from vendors or other sources. By more efficient use of that part of their incomes they could in fact be given adequate good supplies at the same cost. The very large demand that domestic water makes on the life of an individual African may be compared with the small cost, viewed as a proportion of income, to someone with ample good piped water in a developed country.

Fig 6 analyses the costs in greater detail, showing the cost per litre and per head per day. This brings out that the people who are getting their water from unpiped urban areas usually find themselves paying more than the rich per litre and also more than most other people, so the urban populations are particularly at risk.

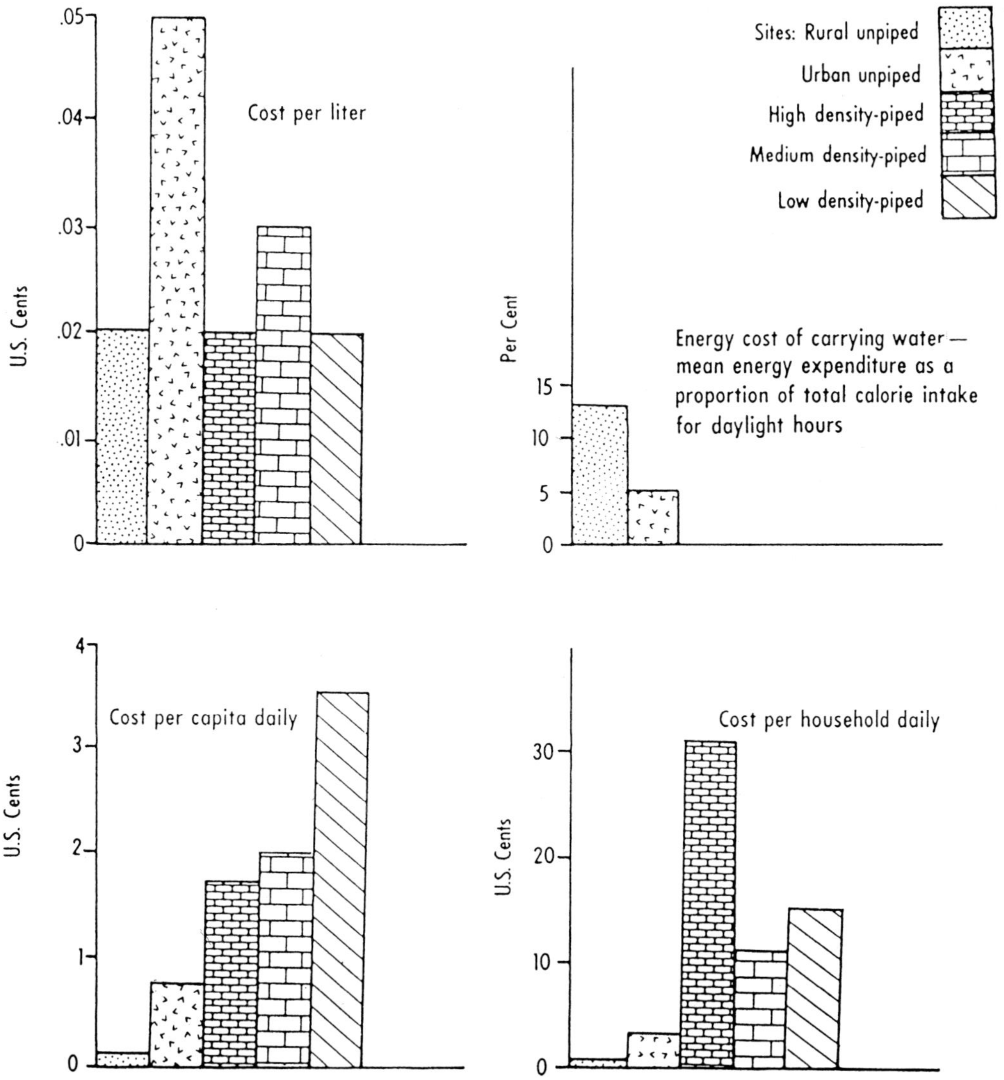

Fig. 6. Cost of water in East Africa by type of site

Water and Disease

Some anthropological papers at this conference showed the range of methods people use to achieve some degree of improvement in their water quality, so there is a very complex, indigenous knowledge of how to improve things, some of which corresponds with what we could consider in medical terms good methods.

It became clear in explaining this, and I think this is crucial to interdisciplinary work, that if you take the names of all the parasites and pathogens and start talking to engineers or politicians their eyes (quite rightly!) glaze over. We were therefore forced to say, "Let's scrap all this biology and let's go back and look in terms of interventions, how we can classify the transmission of water related diseases".

We separated out four different categories. One consists of *water-borne* diseases, that is, where the organism that makes you sick is actually carried in the water. Such diseases include cholera and sometimes typhoid. Often that relationship to water tends to be treated as the only type of health problem related to water: it is not. There is the very important category of transmission, which we call *water-washed*, due to the lack of adequate water for personal cleanliness. It has been shown that many of the diarrhoeal diseases are not primarily getting around in dirty water, they are getting from person to person because people do not have adequate quantities of water supplied. The Minister of Health in his address to us emphasised that the Kenyan policy was to supply adequate supplies of good quality water, in other words addressing both these different issues: both quality of water and availability of it nearby so that people can use it. These two water improvements help reduce a whole lot of diseases. Then there are the *water-based* diseases that are transmitted by organisms that live in the water, of which schistosomiasis spread by water snails is the most important within Kenya. And then surface water generally, just because it is around. If a lot of water is provided but there is inadequate removal of waste water there will be more standing water and some of it may lead to the breeding of mosquitoes and other *water-related insect vectors*. So we need to think about all four categories and not just about the first, which is the only one mentioned in traditional English text books for engineers.

One needs to have a broader perspective, which people within this part of the world now very largely do have, and to relate the different diseases in medical terms to the sort of water improvements that are needed, whether one needs better quality water or a greater volume, and the chances one has of getting rid of the disease. I draw your attention to one example, the only disease we thought could be got rid of by improving water, to remove it from the world, and there is indeed an attempt being made to eradicate guinea worm from the world. It has been eradicated from much of East Africa but there is still some in Uganda. The guinea-worm is transmitted through a minute aquatic crustacean, related to the "water-flea", and simple modifications of the curbs round wells to prevent the worm larvae from reaching the water will halt transmission. There are also other methods. People tend to forget that skin diseases, although they don't kill people, make people extremely miserable and they are very closely related to the availability of water for personal cleanliness. Also, if we look at many of the diseases in a more complicated way, we can analyse to some degree the effects of increasing the volume of water available to people or of improving quality.

Controlling some diseases such as skin infections depends on *more* water, some disease reductions probably depend on solely *better quality* water, and some depend on improving both. The nature of the most appropriate technical improvement to water supplies will vary very much with different environments. The problem in the past was to persuade engineers that there were more options than either providing some sort of a pump at one extreme and at the other extreme putting multiple taps into people's houses with lots of hot and cold water which kept the engineers professionally happy but which meant that only very few people benefited. We concluded from our studies that there was a great range of options for gradual improvement. One can't go from one extreme to another in a single jump at low cost, but in almost any community, by taking a lot of thought and having a flexible engineer who thinks very carefully about how best to do things, it is possible to make some improvement. Maybe one can put in a rural pipeline but not alter the quality. Perhaps the situation in peri-urban areas needs something done about quality as well, but there are a whole lot of intermediates between no improvement and the sort of house that we would all ideally aspire to. Some of these, such as a single flow regulated tap per household, can bring the volume used up to about 40 litres, the sort of level at which most of the health benefits are available even if supply is inadequate to wash your car or water your garden. A level of quality may be attained where most epidemics will not occur. The cost of these different improvements will vary in different parts of East Africa as will the benefits. To get the best benefits in terms of the most health for the least cost will require different improvements in different areas.

Sanitation and Disease

We may now move to look at the other end, that is to say sanitation. We have learned from other papers about the hazards from faeces transmitting infections and again, depending on the infection, what happens between faeces being passed and passing on infections. Some helminth eggs have a latent period, that is to say they are present in faeces if you have got an infection but they won't infect anyone else until they have sat around in the soil for some time. Some eggs will rapidly cease to be infective, some will infect other animals and so multiply in the environment. So some infections in the faeces will drop off over a few hours in the environment whereas others, in the case of the geohelminths, are not infectious straight away but after they have sat around for some days and weeks they will become infectious and stay infectious for months and years. In other words, as with water, there are different ways in which we need to worry. Short survival time eggs and cysts are transmitted by the direct route, that is to say infection throughout a family as a result of dirty hands and the like, so that personal cleanliness and personal hygiene are the crucial things in preventing transmission. There are other infections that depend on keeping the site of the latrine clean and then there are others that will be prevented by disposing of the compost, the material that has been sitting in the latrine for a long time, in a safe way. So again there is diversity.

The vector borne infections such as the schistosomes, as is the case with most parasitic infections, have a very focal pattern of transmission, that is to say that there are just a few places where there is very close contact between people and infected snails and water and those are the places upon which to focus emphasis. There are often annual cycles in the transmission. Sometimes the pond full of snails, which is very good for transmitting schistosomes, dries up for part of the year. Also behavioural issues are important. One small pond was by far the most important in an area for the transmission of schistosomiasis. A picture of it with a little girl provides an answer. She is wearing her school uniform and the path from the school to the village runs past the pond. So children push each other into the pond numerous times when there is water there and they play in the water, whereas elsewhere the risk is much lower.

Behaviour and environmental change are closely linked both in transmission and in environmental control. Returning to guinea worm and its eradication, a very simple alteration for putting an outward sloping curve on a well means that spilt water drains on to the land and dries, and does not get back into the well to transfer the guinea worm back into the water where it will infect other people after going through the intermediate host.

Environmental Change

There is not time to address in detail the many effects that environmental changes will have on malaria. Larger scale, global environmental changes occur and also have possible effects. Climatic changes, especially global warming, may lead to increased malaria in hilly areas, as in Kenya, Ethiopia and Rwanda, where the feet of the hills are in highly malarious countries and the peaks of the hills are way outside the malaria transmission zone. It is there that if there is a significant degree of global warming we shall notice changes in the distribution of malaria. In fact environments are changing all the time. The old idea that they were static is just not tenable. Sometimes the changes are in environmental components, such as the use of insecticides, which may also affect people. We are quite good at knowing what these changes do to diseases and how to control them, given resources. We are much less good with things that change the whole of our life systems. The green revolution in Asia not only improved the yields of rice but totally changed the way in which farming is done. If people move from a rural area to an urban area their whole life changes and we can't just look at single bits, we have got to look at all the changes together. And if we change the habitat, by for example, water resource developments and irrigation we shall again produce very complex changes in health patterns.

I wish now to consider some issues particularly related to urbanization because our conference is particularly focused on urbanization and its consequences for those living in peri-urban areas. In the past, a particular problem with urban development has largely been regulation towards over-design; that is to say, the standards required for urban development, housing and water were so high that in practice they couldn't be achieved by relatively poor people. Consequently the issues were avoided by either bribery and corruption of the officials who should enforce them or in other unsatisfactory ways. There was then a move towards "sites and services:" providing a minimum for people moving to urban areas to give them new sites with the basic services available, where they could build their houses. But this too has proved quite expensive and it is clear that in the present general climate of economic pressures the idea of payment for water has become much more acceptable than it was in the past.

Our problem, as in relation to so many other issues, deals with what is the right public-private mix of funding. I am far from clear that that has been worked out satisfactorily yet. I am quite sure that the extreme views sometimes propounded by my own country favouring the extreme preponderance of the private sector may well be as far from the mark as those that would put the entire onus upon the public sector. Certainly an important local and national debate for the next few years will be getting the right balance between the role of the state and the role of private supply. We see in relation to water kiosks an example where the right management of this can be extremely effective. As set out in a paper from our colleague this morning who had been working in the urban situations in Kenya, where a community water supply was developed through a kiosk system, it was possible to provide water at a fifth of the price of the completely privatised kiosks. In the latter, those who were managing them were clearly successful and very profitable in their activities but the poor were suffering. By carefully planning and exploring the regulatory mechanisms needed to control either cooperative or private sector activities it is often possible to move forward; some degree of community participation is needed for sustainable improvements.

I wish to make some general points about the broader view of environmental change. There are large dams and irrigation projects being developed throughout Africa, among them the Volga Dam, a huge water development project in West Africa. The dam had many complex consequences: people were displaced, the trees rotted; that increased the insects eating the trees, fish increased enormously, vast numbers of fishermen moved in and they were already infected with *Schistomosoma haematobium* and there was a very large epidemic of schistosomiasis. In addition to the complexity of the changes is their long duration. There have been continuous changes occurring in the biology of this system for over fifteen years and it is only now settling down. One has to think very long term in planning what will happen. The population movements that occur are very complex and each population has different health and sanitation problems.

Small ponds and water developments raise quite different problems. For a very small dam, again man-made, the builder hasn't borrowed money from anywhere, no-one has done an environmental impact assessment and nobody has thought about the health issues. But if the problems were sorted out for a small dam, for example in Nigeria, there are between 500 and 2000 similar little dams being constructed so that any answers found to small problem situations can be replicated and so it is possible to systematize improvements.

Parts of South-East Asia illustrate the extreme complexity of forest malaria where environmental changes affect not only the ecology of transmission but the politics of what goes on at the boundaries, since many people depend on smuggling and do that at night when they get bitten by the mosquitoes. They may undertake illegal mining and create ponds where the mosquitoes breed and so on.

So one gets with environmental change very complex results, and in the case of the major vector-borne diseases which affect places such as Kenya, it is interesting to see how approaches to control that use pesticides, molluscicides and chemotherapy, come and go over a period of decades. At the moment we rely very heavily, and rightly so, on chemotherapy. But the one method that vectors such as snails don't find an answer to are really good environmental management methods. The same is true for malaria where we had total commitment to insecticides some time ago, then to chemotherapy but in the case of both of these we are running into enormous problems. Environmental methods have a limited application for malaria control in Kenya. The point which is very relevant for our anthropological colleagues, and the collaborations which are demonstrated at this meeting, is that chemical vector control or chemotherapy make relatively few demands on people to change their behaviour, whereas environmental methods of disease control do require substantial input into behavioural change.

We should aim not only to prevent development projects damaging the environment and making people's health worse, but we should actually see when we are building a project, an irrigation scheme or dam, how we can use that to *improve* people's health, how we can use the fact that we are managing water to ensure that we also improve domestic water supplies and facilities available to people. So there is interaction of health policy and development policy.

At the beginning of this century the argument for better health was because it was felt that socio-economic development was impossible without better health. The tough-minded profiteering capitalists and colonialists in India wanted better health for the people that work on the tea estates because they reckoned they would not get enough tea to make a profit if they didn't. We then had a period of about the last 30 years when people rather viewed development as the means to health, and health was merely a welfare good. It is good to see that with the World Bank's recent report on health they have gone back to realising that health is an investment. Money put into the Ministry of Health is not money put just into welfare but also an investment which will enable people to develop.

On the technical side the number of advances that we shall see is probably not vast. But it is interesting to realise that 20 years ago it was a great struggle to get people to take waste stabilisation ponds seriously, but they worked very well in the tropics: they don't work so well in cold countries, and it was therefore difficult to persuade people who had been trained in cold countries that this was a good idea. In fact it is an extremely efficient method of dealing with human wastes; there are various other relatively unsophisticated but very important technical advances or technical methods which are highly appropriate to particular situations in a country like Kenya.

There is also, for the future, a need to focus on the relationship between household and community. There was a time when the World Health Organization paid great attention to primary health care for patients and for community action. Communities are very complicated as you all know and there are several levels: the district level, the household level, as well as the village. When we are thinking in terms of peri-urban developments there are several levels: the household, the neighbourhood and the municipal health authority where attention is needed. There is also a need to get the right balance between what I would call a public health approach, trying to improve the situation for everyone, for example improving water supplies for everyone or providing maternal and child health services for everyone, to reduce the number of people who are at risk. The alternatives are targeted at-risk approaches which just focus on a few people who are thought to be most at risk. It is hard to achieve the right balance between these different approaches in choosing interventions, partly because it is difficult to bring together the data from many different disciplines in order to really understand the spread of diseases and how they are to be prevented.

Peri-urban Problems

I have already mentioned the problems of the urban fringe where there is a combination of immigration, poverty, overcrowding, a lack of facilities, poor housing and the accumulation of water. Often more water comes in than is drained away properly. The complexity of this is not dealt with in detail here, because it is what most of the conference will be discussing, to try and find ways forward in this very difficult situation. A feature of the edge of African cities is how they combine the rural and the urban. This makes some parts of African cities so attractive, as they have the virtues of urban and of the rural living, but they also have the problems of both. As people move from a low to a higher income and move to a city they undergo demographic changes, what is called the demographic transition, where the population moves to a lower child mortality and a lower reproductive rate. At the same time the population exchanges the diseases of youth, the infections, for the lifestyle diseases: coronary disease, cancer and other illnesses that characterize an ageing population. The problem for the periurban poor is that they are in danger of getting the worst rather than the best of this transition: they get both infections and life-style illnesses which makes for a difficult but hopefully transient situation.

Those who come to help them with projects of development aid may seek nice tidy short-term projects that they can complete, monitor, evaluate and then go away and feel happy. But it is crucial that those projects be integrated into the sector as a whole, that projects which tend to be capital investments move much more towards programmes where the emphasis is more on maintenance and recurrent budget aspects. Those of you who read World Bank literature will have seen diagrams of a project cycle which goes through a tight circle. It is one of the most misleading political documents (by accident) that one could have. The Bank is happy when something is built; it has funded something and then goes away and that seems fine, but what is important, especially for water and sanitation, is what happens over the next 20 or 30 years. The unglamorous aspects, that is to say operation, maintenance and monitoring of what is happening, are the things that make the difference between good and bad water supplies. The situation is that resources are scarce but on the whole capital resources are less rare than recurrent budget ones.

There is another series of topics which tend to dominate discussions and which in many ways could be described as seesaw items. For example the balance between centralization and devolvement locally. These are debates that take place every 10 or 20 years. At intervals we decide that centralization is the answer to all problems of local government. After some years of that we decide that decentralization is the answer. Similarly there is a discussion between what should be provided and what should be self-help. If one stays in a field for 30 years, each alternate position is rediscovered at least once if not twice! I think very broad slogans of that sort have a limited value. What is crucial is to try and work out *what* is best decentralized and what it is appropriate to keep centrally. In a particular situation, if you are trying to get villages or small groups of people to do things for themselves, and a great emphasis is on self-help, it is crucially important that they have very good technical advice available when they need it, so that they don't waste their time on something which has previously been shown not to work. In this brief paper I have not done adequate justice to the role of the community. Community participation should be not just in doing things but in deciding what it is that shall be done. This has already been addressed in the conference and undoubtedly there will be more on it.

On the whole, where water and sanitation improvements have been successful a large range of different inputs have been needed. It has required a methodology for analyzing the problems, it has required effective intervention which means paying attention to general health as well as water and sanitation but it is a question of choosing what is the most appropriate for the particular situation. It has required a clear perception of what the issues are. Quite often, through misperceiving the issues, effort is put in the wrong direction. If all effort is put towards quality and none into availability of water you do not get all the health benefits.

Effective action requires strong political commitment. It is not possible for the technocrats to achieve anything lasting unless there is a strong political commitment and involvement of the community at all levels. It is very gratifying that people have stopped thinking in quite such compartmentalized way as 70 years ago. A doctor now hopefully doesn't just think about patients but also about the community. The engineer no longer thinks only about pipes. Dubos wrote several fine books and he was very worried. He wrote that one of the terribly dangerous things is that man is so adaptable. People can actually adapt to living on the pavements in Calcutta and coping with terrible slum conditions. In one sense that is amazing, but that is not the way that people should have to live. We should try to design environments which will be suitable to people's development needs. This conference is looking at some of the ways in which that can be done in relation to water and sanitation. I know that the honourable Ministers of Health and of Environment are concerned with the much broader context and that the rest of the conference will continue the broad approach set by them in their opening words. Let us hope that this wide understanding will also be reflected in the development of all African peri-urban areas.

THEME I : WE NEVER KNOW ENOUGH

Chairman: Prof. D.M. Otieno

Paper 1. COOPERATIVE EFFORTS AND CONFLICTS IN COMPOUNDS AND NEIGHBOURHOODS TO RID THE AREA OF STANDING WATER

Nelson Waweru, Deputy Chief Public Health Officer,
Ministry of Health, Kenya

Introduction

Standing waters are a major cause of concern to public health officials as well as enlightened members of the community because of the negative public health implications associated with them. Unfortunately the dangers are not readily understood as many of our people do not appreciate the relationship between standing water, mosquito breeding, and malaria or other water-related diseases.

Consequently, calling for community action to drain stagnant water does not receive the importance or priority it deserves and creates conflicts which would not otherwise have arisen. Strategies which can facilitate action at individual and community level are urgently required as the problem of standing water in periurban settlement cannot be left to the local authorities or to the Government.

Health Problems Associated with Standing Waters

Standing water can create a good environment for the breeding of water-based and water-related disease vectors. Contamination by excreta can also lead to a favourable environment for the transmission and spread of water-borne diseases. While malaria, schistosomiasis and typhoid are diseases of major concern, standing waters create unhealthy environments arising from dampness and smell nuisances arising from decomposing wastes dumped in such water.

Sources of Standing Water

There are various sources of standing water which occur as a result of human activity or natural causes. The most common ones are:

- ◊ Storm water pools after heavy rains
- ◊ Burst water pipes
- ◊ Burst sewerage pipes
- ◊ Blocked sewer pipes
- ◊ Poorly maintained house drainage
- ◊ Poor siting of buildings and unsuitable grounds
- ◊ Irrigation activities
- ◊ Construction works without reinstatement
- ◊ Mining and quarrying activities
- ◊ Flooding of areas next to large rivers
- ◊ Poor disposal of containers
- ◊ Abandoned pit latrines, septic tanks and cesspools

Conflicts at Compound and Neighbourhood Level

Standing water is viewed by the community as a nuisance at both compound and neighbourhood level depending on the location. The main areas of conflict arise from :

(a) Physical planning problems. One major aspect is the development of unplanned residential, industrial and commercial areas without taking into account the attendant drainage problems. This creates difficulties in the control and disposal of liquid wastes resulting in various health hazards. Under these circumstances there is a conflict between the residents, the local authority and the landlords.

(b) Conflict of land use. The land in periurban areas is usually free hold and the local authority has no direct control over it. The individual developer does what he likes without caring for the impact of his activities on the immediate neighbours.

(c) Development activities which do not adhere to the principles of zoning in order to separate residential, commercial and industrial areas result in conflicts especially when such developments generate accumulations of liquid wastes.

(d) Land ownership. Most periurban dwellers do not feel responsible for nuisances occurring on public utility land even where they are the authors of such nuisance. This leads to conflicts between them and the local authority.

(e) Local authority services. Conflicts arise when the community is charged for services that are not efficiently rendered by the local authorities. For example, a leaking pipe or blocked sewer may create standing water next to residential areas in which case the community expects the resulting nuisance to be abated by the local authority.

(f) Conflicts between formal and informal settlements. Most informal periurban settlements do not have drainage systems while the neighbouring planned settled areas are catered for. This sometimes leads to stagnation of water from the informal settlements in the planned settled areas causing conflicts between the two neighbourhoods.

(g) Water use conflicts. Some periurban areas have acute shortages of water for domestic use. Any available water, for example in ponds or abandoned quarries, may be the only source of water for a community and any attempt by whichever authority to have such water drained, leads into con-flicts with the community.

(h) Conflict between tenants. This may occur when water from a neighbour flows and stagnates in another compound.

(i) Conflicts between landlords or between landlords and tenants. This occurs when the landlord is expected to take action and does not act or when the landlord thinks that the tenants are negligent.

Cooperative Efforts to Deal With Standing Water

Experience has shown that there are limited cooperative efforts taking place to deal with standing water issues in periurban areas. The following may be cited as the main constraints hindering cooperation:

(a) Periurban dwellers, despite their close proximity to one another, see themselves as individuals and not as a community: They seem to have no time to identify their common problems and interests.

(b) Lack of knowledge on the need to drain standing water on the part of the community.

(c) Lack of community sensitization and mobilization to generate cooperation and active participation.

(d) Lack of coordination by the main actors dealing with issues related to standing water and its drainage.

(e) Inadequate resources to deal with the problems of standing waters, especially when large water bodies are involved.

Control Measures

There are various methods of approach to the problems of standing waters depending on prevailing circumstances. At the compound and neighbourhood level the methods of dealing with standing waters are generally within the inhabitants capacities since very simple methods are employed. However, for the community to act, there is need for concerted efforts to remove the constraints stated above.

The Public Health Act (Cap. 242) and the Mosquito Control Act (Cap. 246) of the Laws Of Kenya are very elaborate and specific on the measures to be taken against anybody creating a situation that may lead to standing waters (see paper 15). It should be noted that the two acts are categorical in their approach which may undermine the desired cooperation from the community. It is therefore advisable to use the law as a last resort. However, in certain circumstances the application of the law may be the only appropriate approach.

Conclusion

In this presentation I have tried to highlight the problems of standing water in periurban settlements as well as the constraints encountered while looking for possible solutions. Good health being the responsibility of every individual; there is need to educate, mobilize and guide the periurban communities to collectively formulate strategies to prevent diseases and enhance living in a healthy environment. The Primary Health Care concept seems to me to be the most appropriate approach as local authorities do not have the means to adequately deal with the problem.

Paper 2. BIOLOGICAL CONTROL OF URBAN AND RURAL FILTH FLIES

Moses Makayoto, International Centre of Insect Physiology and Ecology (ICIPE), P.O. Box 30772, Nairobi.

The Use of Biological Agents for Control of Insects

The last few decades have been characterized by the use of large amounts of chemical insecticides for control of crop pests and vectors of diseases for both human and livestock. The disadvantages of continual use of chemicals, however, have caused concern worldwide because of their adverse effects; their broad attack on other biodiversity, the pollution of the environment due to their accumulation in the soil, water, air, agricultural products and animals, resistance development by target insects, and the high costs which are not within the means of most people of the developing world. Consequently research and development of other alternative pest control methods have been initiated. One such approach which has received a lot of attention is the development of environmentally safe and sustainable technologies called Biological Control. This has involved the use of insect parasites and predators, and entomopathogenic microorganisms such as viruses, bacteria, fungi and protozoa. To date the bacterium *Bacillus thuringiensis (Bt)* is the most widely used microbiological biopesticide. This has been due to its efficacy against intended target insects, its favourable biosafety aspect as far as humans, other vertebrates and beneficial organisms are concerned, and the commercial feasibility for its production.

Filth Flies: Epidemiology and Associated Diseases

The existence of the filth dipteran flies and their evolution over the years has taken place in environments close to man and other animals. They are usually found in areas such as cattleenclosures, garbage areas, pit latrines, houserefuse, sewage systems and on widely disposed human waste in places without any sanitary systems, decaying animals and vegetable matter. In a detailed literature survey of general filth flies associated with human and animal diseases, their biology, distribution and management (3), a number of important points were noted.

In terms of transmission and dissemination of disease pathogens to both man and animals, filth flies are known to mechanically transport on their mouth parts, through their vomitus or faeces either to human and animal food/feed or directly to hosts (note 1). This is due to their feeding on both human and animal wastes such as faeces, sputum, nasal secretions, pus and urine. The adults of some of these flies are non-parasitic but have been known to be responsible for bacterial, chlamydial, protozoal, helminth and viral infections.

In certain cases, larval stages of some species live parasitically within human and/or animal bodies, thereby causing diseases (myiasis). In general, globally and Kenya-wide, the common diseases caused are trachoma, conjunctivitis, bacillary dysentery (shigellosis), cholera, typhoid fever, colventeritis, amoebiasis, myiasis, mastitis and thelaziosis. Other genera such as *Musca, Stomoxys* and *Lyperosia* are commonly found around nostrils of man and animals. In large numbers, flies are a nuisance and cause irritation and in the case of domestic animals lead to interruption of normal grazing/ feeding, thus causing a severe drop in weight gain and milk yield.

In their metamorphosis, most of the flies go through egg, pupal and adult stages. The habitats selected for eggs deposition is dependent upon species, season and geographical location.

Intervention Method at ICIPE: *Bacillus thuringiensis* (Bt) R and D

For the last few years, ICIPE has been involved in the development of *Bacillus thuringiensis* (*Bt*) based biopesticides for control of its target insects including, among others, medical vectors such as mosquitos, sand flies, filth flies and tsetse flies. The areas of research comprise isolation, bioassay, biosafety, field testing and optimization of growth conditions. A pilot study for evaluating the feasibility of using *Bacillus thuringiensis* as a biological insecticide against flies in tropical countries has been undertaken, covering field testing in pit-toilets, garbage collection and slaughter houses (condemned meat-pits), socio-economic and epidemiological studies, optimizing fermentation (production) conditions using locally available raw materials such as soya-bean, cow pea, molasses and process design and economics as pertains to the tropical situation. Details can be consulted in a number of publications (1,2,3,4,5).

Findings: Relative Abundance of Filth Flies, Breeding Sites, and Insecticides

Studies were carried out in Kibera slums (Nairobi) and the small towns of Nguruman (Kajiado district) and Marigat (Baringo district) before the intervention (3). In Kibera, genus *Chrysomyia* was predominant (61%) followed by *Musca domestica* (37%). Very few flies in the genera *Lucilia* and *Sarcophaga* were found. In both Nguruman and Marigat areas, *Musca domestica* was the most predominant, 93% and 72% respectivly. Other fly species in the two areas were *Musca sorbens*, *Haematobia minuta* and *C. chloropyga.*

Pit latrines, garbage dumping sites and filth drainage systems were common breeding places in Kibera while in Nguruman and Marigat, cattle and sheep/goat enclosures and heaps of manure were common. The filth fly borne-diseases were mainly trachoma, conjunctivitis and myiasis.

The widely used method is spraying and dosing of chemicals in both rural and urban areas of Kenya. The study in Nairobi established that the main insecticides used during major outbreaks are Dichlorus E.C. (200 litres), Fenitrothion 50% E.C. (100 litres), Propux (330 kg), Pyrethrum 25% solution (100 litres), Phenthoate (600 litres), Lindane 20% E.C. (100 litres) and aerosols (56.8 kg). In the slum areas the filling up of pit latrines with faeces and the subsequent crawling of the maggots all over the place led people to using all sorts of methods which they think work, which include putting ash, used batteries etc. into the toilets. Fly-traps are also occasionally used.

Mode of Action of *Bacillus thuringiensis*

Bacillus thuringiensis popularly referred to as *Bt*, is a Gram-positive and spore-forming soil bacterium characterized by its production of crystalline inclusion proteins (endotoxins) during sporulation. These pro-teins have proven pathogenic to the larval stages of certain insects. Some serotypes of *Bt* produce both endotoxins and exotoxins. The many different *Bt* strains available exhibit toxicity to a number of insects, including flies, crop pests as *C. partellus, B. fusca, S. calamistis* and other dipterans such as mosquitos and black flies. This report is mainly on *Bt* serotype H1 which is effective against fly larvae.

The bacterium during growth produces an exotoxin referred to as thuringensin. The toxin affects the moulting and pupation of the fly larvae and causes either mortality or deformation of the larvae or adult fly depending on the dosage level. It is prepared in three forms; a broth of the bacteria, spores and the thuringensin; a concentrated paste of spores and bacteria or a spray dried powder containing spores and bacteria.

Production and Cost of Dudustop and its Efficacy

ICIPE has established conditions for production of Dudustop using *Bt* serotype H1 (fly factor). It is carried out in a laboratory fermenter (14 litre) using soya-bean (3%), molasses (1%), and phosphates (0.5%). The cells and spores average 10^9 per ml. The centrifuged paste that is commonly used has spores and bacteria as the active ingredients.

The constraint that faces Africa is the lack of production facilities to produce large quantities with strict sterility to avoid contamination. These facilities are readily available in the developed world. However, it is likely to be cheaper to produce *Bt* in Africa using locally available raw materials than to import in bulk.

Ingredients of Dudustop include bacteria, spores and thuringensin. Depending on whether the formulation is broth, paste or powder, a dilution factor of the paste material of 1:40 with water has been shown to be very effective and this reduces costs considerably. Each latrine is dosed with one litre of the diluted material. Apart from pit latrines Dudustop has been shown to control flies in condemned meat pits and garbage heaps. The effect is seen in between 5-10 days and within 3-4 weeks no larvae or maggots are seen in the toilets. Also a considerable decrease in the population of adult flies takes place.

It is envisaged that with its long lasting effect and the dosing frequency of twice a year, the treatment cost per toilet per year would be in the region of one to two USD. This is definitely much cheaper than frequent spraying with chemicals.

Dudustop has been applied in various areas in Kenya including the Kibera slum in Nairobi, Kakamega, Bungoma, Busia, South Nyanza, and North Eastern Province at Dadaab refugee camp (5). The application has now been extended to refugee camps in the Gambele area of Ethiopia and Ngara, Tanzania.

Biosafety and Registration of Dudustop

Numerous safety tests on *Bt* have been carried out over the last two decades. The tests involved inhalation, percutaneous and intravenous toxicities, allergenicity, long-term feeding tests and random exposure to man. Such tests of safety and efficacy are required by regulation before a product is registered.

Dudustop (whose active ingredient is the thuringensin) has undergone testing for both toxicological and likely mutagenic properties. Sub-mammalian tests (10,12) have been done with the fruit fly, *Drosophila melanogaster*, root tips of onion *Allium cepa*, and yeast, *Saccharomyces cerevisiae*. Mammalian tests (6,7,8,9,11) have also been done using mice, rats and human lymphocyte cultures for toxicities and also Guinea pigs for allergenicity tests (6). No inhalation, allergenicity and toxic effects were noticed on the species tested.

In mutagenicity tests of *Bt* the Ames test (12) showed no mutagenicity. It is important to note that enzymatically dephosphorylated exotoxin is biologically inactive and dephosphorylating enzymes in the human digestive tract also inactivate the toxin; this phenomenon also occurs in a number of insects (11).

Dudustop is now accepted as a trade name by the Kenya Industrial Property Organisation. On registration, details pertaining to it, are now with the Pests Control Products Board for commercial application.

Conclusion and Recommendations

Dudustop has been shown to work very well in Kenya, Tanzania and Ethiopia. It is therefore apparent that Dudustop application can be widely used in these areas and in other tropical countries.

The wide application of Dudustop would warrant a detailed study on disease impact assessment.

The following conclusions and recommendations were recognized:

- ◊ Socio-economic studies have indicated that low-income groups can afford Dudustop
- ◊ Epidemiological surveys have justified an intervention package.
- ◊ Technical studies show that a production facility can be set up in Africa. Requests from UNHCR, OXFAM, and CARE indicates that Dudustop is in demand in many refugee camps.

Note 1. In terms of their association with filthy places the flies in the genera *Musca, Calliphora, Lucilia, Muscina, Fannia, Sarcophaga, Stomoxys, Chrysomyia. Callitroga, Ophyra, Cordylobia, Synthesiomyia, Atherigona, Anthomyia, Auchmeromyia, Hydrotae, Haemotobia, Eristalsis, Gymnodia* and *Lyperosia* have been recognized.

References

1. Carlberg G. (1986). *Bacillus thuringiensis* and Microbial Control of Flies. *MIRCEN Journal*, 2, 267-274.
2. Carlberg, G.et al. (1985). Microbial Control of Flies in Dar es Salaam with a *Bacillus thuringiensis* Preparation, Mucabac. *MIRCEN Journal*, 1, 33-34.
3. ICIPE, Kenya and Provio Oy, Finland (1990). Pilot Study for Evaluating the Feasibility of Utilizing *Bacillus thuringiensis* as a Biological Insecticide against Flies in Tropical Countries. Final Report.
4. Carlberg, G.et al. (1991). Control of Fly larvae with *Bacillus thuringiensis*: Field and Socio-economic Studies in Kenya. Paper presented at the International Conference on *Bt*, July 28-31, Oxford, London.
5. Carlberg, G. and M. Makayoto (1993). Microbial Control of Flies with *Bacillus thuringiensis*. Field tests in the Refugee Camp in Dadaab, Kenya. Final report to the Academy of Finland.
6. Carlberg, G. (1973). Biological Effects of the thermostable Exotoxin Produced by the Different Serotypes of *Bacillus thuringiensis*. Reports from the Department of Microbiology, University of Helsinki.
7. Meretoja, T. and G. Carlberg (1977). The Effect of *Bacillus thuringiensis* and of Cell-free Supernatants of Some Other Bacteria on the Mitotic Activity of Human Lymphocytes. *FEMS Microbiology Letters* 2, 109-111.
8. Meretoja, T. et al. (1977). Mutagenicity of *Bacillus thuringiensis* Exotoxin Mammalian Tests. *Hereditas*, 85.
9. Kahkonen, M. et al. (1979). Mutagenicity of *Bacillus thuringiensis* Exotoxin III. Sister Chromate Exchange in Rats in *vivo*. *Hereditas*, 85.
10. Linnaima, K. et al. (1977). Mutagenicity of *Bacillus thuringiensis* Exotoxin II Sub-Mammalian Tests. *Hereditas*, 85
11. De Barjac, H. and M-M Lecadet (1975). Elimination de Toxine Thermostable de *Bacillus thuringiensis* Apres Ingestion chez la souris C.R. *Acad. Sci. Paris*, 280.
12. VTT Food Research Laboratory (1993). The Mutagenicity Testing of Samples of *Bacillus thuringiensis* Using the Ames Test. Research Report No. EL11287/93.

Paper 3. CONGESTION AND HUMAN RESPONSES TO GETTING RID OF WASTE

Julia Kunguru and Mukami Kariuki, World Bank Regional Water and Sanitation Group, P.O.Box 30577, Nairobi, Kenya

Introduction

The search for improved quality of life has resulted in growing migration to urban areas. As a result, the population of most urban centres in Kenya has rapidly grown. However, as in-migrants arrive in urban areas, many find that they can only afford the cheap rental housing in informal settlements.

Informal settlements typically consist of semi-permanent structures, built in an unplanned manner on legally or illegally occupied land. These settlements are a response to a regulatory environment that has restricted the supply of affordable land and housing.

Often located on environmentally fragile lands, many informal settlements lack basic infrastructure and services. Environmental sanitation problems in these settlements are extensive and have a significant negative health impact on residents.

The following are typical characteristics of informal settlements identified in an inventory of Nairobi's informal settlements (1):

(a) Owners of structures have either a quasi-legal right of occupation or no rights at all.

(b) Structures (houses) are constructed primarily of temporary materials and do not conform to minimum standards.

(c) The majority of structures are let on a room by room basis and the majority of households occupy a single room or part of a room.

(d) Densities are high, typically 250 units per hectare compared to 25 per hectare in middle income areas.

(e) Physical layouts are relatively haphazard making it difficult to introduce roads, pathways, drainage, water and sanitation.

(f) The majority of inhabitants have low or very low incomes.

(g) Urban services such as water and sanitation are non-existent or minimal .

(h) Morbidity and mortality caused by diseases stemming from environmental conditions are significantly higher than in other areas of the city - "owing to poor sanitation, lack of potable water, poor drainage, uncollected refuse and overcrowding" (2).

This paper analyzes and updates information provided in two case studies on environmental sanitation in the two informal settlements of Kibera (Nairobi) and Maina (outside Nyahururu town centre). The case studies were carried out by the UNDP/World Bank Water and Sanitation Program - East and Southern Africa in 1992 and 1993. The views expressed in this paper are the opinions of the authors.

Overview of the Study Areas

Background to Kibera: Kibera is the largest informal settlement with an estimated population of over 400,000 people living on an area of 110 hectares located south-east of Nairobi within the city boundaries seven kilometres from the central business district. The informal settlement comprises 10 villages with 20,000 to 50,000 persons in each village.

Table 1: Population and Densities in Kibera, by Village

Village	Area in hectares	Average no. occupants per room	Population Density Persons/ha	Total population
Laini Saba	16.5	4.5	1,656	27,340
Soweto	20.6	5.1	1,840	37,949
Mashimoni	12.0	5.1	1,945	23,437
Gatuikera	25.6	5.2	2,042	52,234
Kisumu Ndogo	21.9	5.0	2,203	48,340
Lindi	25.9	4.8	2,297	57,715
Makina	39.5	5.3	2,423	95,636
Silanga	21.3	4.9	2,527	53,850
Kianda	18.3	5.4	3,895	71,366
All Villages	**22.4**	**5.0**	**2,314**	**467,857**

Uncertainty about the future of the settlement continoues to dominate the area. However, despite this, the settlement's population has continued to grow. The 10th village known as Raila has been established since the initial case study was prepared.

The distribution of population by gender is biased toward males (64%) (3). This is probably a result of the high percentage of single male residents who are seasonal labourers in the nearby industries. The majority of residents are aged between 20-29 years, however, there are large numbers of children of school age many of whom are not enrolled in school. Some of these are street children and many are from female headed families.

Basic infrastructure in Kibera: Kibera lacks adequate infrastructure facilities such as water, roads, sanitation and solid waste collection facilities. In addition, the area is crowded with temporary, congested residential and commercial structures. The following provides an update on the current environmental sanitation in Kibera.

Because of the unplanned nature of the settlement, access to water is limited, sometimes being provided on an intermittent basis by vendors who are disconnected by the Water and Sewerage Department of Nairobi

Table 2: Water Supply Coverage in Kibera: Water points by Village

Village	Total No. of Water Points	Public Water Points	Private Water Points	Average no. people per water point	Water points with storage
Silanga	71	4	67	758	56
Laini Saba	94	3	91	291	67
Makina	121	7	114	790	53
Mashimoni	76	3	73	308	51
Gatuikera	99	5	94	527	48
Kisumu Ndogo	96	8	88	503	45
Soweto	69	3	66	550	62
Lindi	97	3	94	595	46
Kianda	96	6	90	743	40
All Villages	**819**	**42**	**777**	**562**	**468**

Source: Survey of Water and Sanitation Facilities in the Kibera Squatter Settlement, 1994, page 5.
NB. Does not include the new settlement Raila.

City Council for non-payment. Although there are over 800 water vendors connected to the NCC mains (as shown Table 2), due to water shortages, prices are inflated. However as a result of community water kiosks built in the late 1980's with the support of NGOs, price increases have been somewhat controlled (see box 2).

Although the sewer line runs through Kibera, on-site sanitation is the primary means of excreta disposal. Despite efforts by various NGOs to improve on-site sanitation, the situation in Kibera has continued to deteriorate as population pressure and land scarcity have made it extremely difficult to construct new latrines. Once they fill up, some latrine sites are converted for other purposes such as kiosks.

This situation has been exacerbated by the breakdown of the latrine emptying service which ceased operation in 1993. Further discussion on the latrine emptying service is contained in box 1.

Box 1: ***Latrine Emptying Service in Kibera***

In 1990, the charge for latrine emptying service was 150/= per load (approximately US$ 4). Of this 40/= per load was paid to the Nairobi City Council. The charges were based on full cost recovery. Results at the time indicated that there was high demand for the service. The vehicle worked six days a week in order to cover all villages in Kibera. Each village was served on a given day and clients were required to make payments in advance. The initial impact was considerable with about 6,000 plots served in the first few months of 1991 (4).

As a result of inadequate maintenance, the exhauster broke down in 1993. Since the collapse of the latrine emptying service, latrines which fill up are abandoned or covered with another structure eg. a kiosk. The Daily puts it "Already distressed with mud in the wet season, Kibera residents now have to dodge through lakes of human refuse whenever it rains" (5). As a result intestinal parasites such as worms have increased to the extent that the city council has introduced a monthly de-worming programme.

When an NGO named KWAHO (Kenya Water for Health Organization), active in the water and sanitation sector, first aquired the vehicle with funding assistance from NORAD, they did not involve the local administration i.e. district officers and the chiefs. The adminiistration later felt that they should manage the service on behalf of the community and in the process, the community committee was dissolved and another one set up by the chief. The service eventually collapsed as a reuslt of inadequate and poor management.

Solid waste is a major environmental sanitation problem. The city council does not collect solid waste within the settlement, and collection on the outskirts of Kibera is irregular and infrequent. Solid waste is typically dumped in open spaces, such as the railway embankment, and in drainage channels or streams, or burnt. These dumping grounds are also used for defecation by children and adults at night.

With only one main unpaved road in Kibera, which is in very poor condition due to lack of maintenance, the city council has not been able to collect the solid waste from the settlement. Due to increase in population, the road has been encroached on, making it impossible to use. To make matters worse, when it rains the road is rendered totally impassable.

There is no planned drainage system in the settlement. Most drains (which have been constructed through community and NGO efforts) follow natural courses and are unlined. Due to the steep topography of the area erosion and clogging of drains with solid waste is common. All drains in Kibera flow into the Nairobi dam which is suffering from serious pollution.

Due to emptying problems, sullage is no longer disposed of in latrines but is dumped on access roads and drains, resulting in pools of stagnant water which children play in. Often overflow from latrines finds its way into the drainage system. The health risks associated with the kind of conditions cannot be overstated.

Background to Maina village: Maina village was established in 1965 for ex-servicemen of the King's African Rifles. The name, "Maina", is taken from the Kikuyu custom of naming groups of age mates and most inhabitants then belonged to the "Maina" group. The initial residents of Maina were granted 20 years lease which expired in 1985. These were not renewed until 1992, when 99 year leases were granted.

Maina is situated about 2 kilometres West of Nyahururu Town centre. Nyahururu is the largest urban centre in Laikipia District and Maina village accounts for about half of its population of 30,000. Maina village is an informal or unplanned settlement occupying an area of about 1 km^2. It is surrounded by forests and fertile agricultural land. The settlement slopes towards the Gathare stream which is a key water source for Nyahururu Municipality. The low lying area of the settlement close to the stream is swampy and has a high water table. Despite these conditions, in the period since the case study was carried out, a new settlement known as Ukweli has been established in the area along the stream. An estimated 2,000 people now reside in this area and its population is expected to increase as further development occurs.

Basic infrastructure in Maina village: Prior to the Danida funded Sewerage House Connection Project (SHCP), Maina village had limited infrastructure, including a main road, two primary schools, a nursery school and a dispensary.

The village is now served by two water sources: a protected spring and piped water from the municipality. The spring was the main source of water until 1982 when piped water was connected to households in the village through the project. Despite the piped water supply connection programme, many households still rely on the spring which is in poor condition. Disconnections are not uncommon, as landlords frequently fail to pay their water bills.

As a result of a recent failed attempt by the Municipal Council to increase the water supplied by the spring (to accommodate increased population pressure), the community has recently removed the cap from the spring and are using buckets to draw water from it. The spring is the only reliable source of water for the village and has now become a health hazard.

The sewerage house connection programme is non-operational. This is due to blockages caused by misuse (non-flushable materials) and exacerbated by the inadequate piped water supply. Sewage is discharged in the lower part of the village, creating a terrible stench and posing a health hazard. Although the municipal council attributes this to poor design, also inadequate maintenance and hygiene education has contributed to the problem.

In the un-sewered area of the village, latrines constructed through the SCHP are also semi-functional. Due to the high water table, and construction going on in the Ukweli area, many latrines are overflowing. Several plot owners have pulled off the roofs of the latrines for added ventilation, also allowing rain water to fill up the latrines. On some plots, no efforts have been made to build superstructures on the latrines.

The eucalyptus trees which had been planted between the village and Gathare stream (to increase evapo-transpiration) have been uprooted in order to make room for the Ukweli settlement.

The solid waste collection bays constructed through the project are being used by the communities, but much of the waste is dumped outside the bays. Several community members noted that the bays were too high for the women and children who were the main users. The municipal council does not collect solid waste from the village regularly, further discouraging residents from using the bays which were constructed along the main road. Heaps of solid waste were evident along the main road and in the drains running alongside it.

The drainage system constructed by the project has not been maintained and the community health committee notes that the municipal council which collects a service charge from the Maina residents, should be responsible for maintenance. The clean-up activities that were organized by the health committee to clear the drain are no longer held. The result is that the drains are overgrown with grass and have become breeding grounds for mosquitoes.

The National Policy Framework

"The Central Government's position on informal settlements is that squatters should be offered an alternative resettlement site before being evicted. Beyond that there seem to be no government policy dealing with informal settlements" (6).

In the absence of policies tailored towards informal settlements, legislative measures especially the Public Health Act and Grade I and Grade II building codes have provided guidance to local administrations, NGOs and other external support agencies.

The building codes which previously restricted the use of low-cost and semi-permanent materials for construction in urban areas have now been revised by a committee set up by the Ministry of Land and Housing in 1992. However, land tenure and unavailability of land for low-income housing remain the key constraints to the improvement of environmental sanitation in urban areas.

The Role of Support Organizations in Improving Environmental Sanitation

In the absence of an adequate policy framework to facilitate the wholesale upgrading of housing, infrastructure and services in informal settlements, Community Based Organizations, Non Governmental Organizations, and other External Support Agencies have made selective interventions aimed at alleviating poor environmental sanitation and health conditions in informal settlements. The following section describes the role that these agencies have placed in Kibera and Maina Village.

Community based organizations: CBOs have played a key role in facilitating environmental sanitation improvements. They have been instrumental in organizing and mobilizing community members to participate in projects. "In 1992, it was estimated that about 60% of Kibera residents belonged to at least one community organization. The most common organizations are self-help, women and church groups. These groups have a strong community spirit and have worked with various NGOs to enhance access to water" (7).

The most prominent groups in Kibera are the "Afya" (health) groups, registered with the Ministry of Culture and Social Services. They typically consist of 20 permanent residents and landlords, representing each village. The main role played by the Afya group is to organize and mobilize the population in weekly clean up activities in Kibera.

Box 2: ***Community Managed Water Kiosks in Kibera***

The other initiative jointly carried out by KWAHO and the community was the initiation of water tanks and kiosks for community groups and institutions. The community provides labour and materials while KWAHO through WATER AID and UNICEF provides financial and technical assistance. Water charges for water provided by community groups are lower than that provided by private vendors. The community water points charge 1/= per 20 litre container while the private vendors charge between 2 - 5/= depending on the demand and supply for water. Although there are more private water points than community ones, residents prefer the community ones because they are cheaper. Queues at public water points are therefore long.

The major impact from the community initiative has been to stabilize water prices and increase the reliability of water through the use of water tanks since most house-holds do not have storage facilities.

The health committee and women groups feel that their efforts have paid off because they have been able to generate up to 1000/= a month from the sale of water.

A key role played by community groups in Kibera is the management of water kiosks constructed with the help of KWAHO. As illustrated in Box 2, these kiosks have played an important role in reducing the burden of inflationary water prices on the community.

In the case of Maina village, the role played by community groups included mobilization of residents to participate in project activities, regular solid waste clean up campaigns, clearing of drains and proper usage of facilities. More importantly, the community groups in Maina were instrumental in building agreement and consensus on the Sewerage Housing Connection activities proposed by DANIDA. These activities had initially been rejected by the community who were concerned that any interventions could jeopardize their security of tenure. Their land leases had expired several years before the project started and were only renewed in 1994 after the completion of the project.

With the help of KWAHO, and after extensive negotiations with the municipal council the community was assured that their tenure would remain secure. They thereafter committed themselves to facilitating the project by providing land and materials for infrastructure and service improvements. Some even knocked down part of their structures to make way for roads, sewer lines and for sanitation facilities.

Non Governmental Organizations (NGOs): The case of KWAHO. The essence of the NGO approach is not to induce development financially, but to mobilize people into organized structures of voluntary group action for self-reliance and self development. In the case of Maina village and Kibera KWAHO played this role. To facilitate implementation of the sewerage project, DANIDA contracted KWAHO to mobilize the community and train villagers in the importance of health and management of solid waste.

Specific goals included identification of roles and responsibilities of the community in project implementation and management, and assistance in building a structure at the community level that would ensure sustained utilization of the infrastructure with maximum benefits to the people of Maina village. KWAHO was successful in building community interest and participation in the project during its implementation. Community mobilization was phased out during a difficult time when the Government was revisiting the land tenure situation in the area and residents were insecure about their future in the post project period. As a result much of the effort to build community responsibility for improving environmental sanitation has been lost. This is evident from the description of the current situation in Maina presented earlier.Kibera, KWAHO set up a pilot project which aimed at increasing accessibility to sufficient water through construction of kiosks (water selling points) and improvement of ventilated improved pit latrines. 14 women groups participated in the pilot project by providing labour and funds for construction, they also managed the facilities themselves. The success of the water kiosk programme may be attributed to demand for the service which remains evident when one compares prices for water purchased from private vendors with community vendors. The community continues to realize several benefits from these kiosks including lower prices, increased access to water and income for members of the community water vending groups.

While the kiosk program has been largely successful, the latrine project - which resulted in the construction of demonstration latrines in public places and the establishment of a latrine emptying service, has been stopped. This is due in part to land and population pressure which has severely restricted the construction of new latrines. In the case of the latrine emptying service, disagreement over management of the service resulted in its eventual collapse (see Box 1). KWAHO is currently assisting the community to reinstate the service, but requires funding for repairing the vehicle.

From the two case studies, one concludes that KWAHO had the capacity to promote local participation, working with community groups as partners, and emphasizing local self-help initiatives. However, sustainability of community based initiatives in the post project phase remains a complex and difficult activity, particularly in urban communities that are largely tenant oriented and lacking security of tenure.

The Role of the Authorities

In both projects, the role of the central government was to provide a framework within which project activities could be undertaken. Although the government has a stated policy on informal settlements, the legislative framework for infrastructure and service improvements and upgrading in such settlements is somewhat complicated and unclear, particularly in regard to land tenure.

Both the Maina and Kibera projects were agreed and facilitated by the central government through the respective administrative offices and it was also responsible for registering and facilitating the activities of CBOs and NGOs in these projects. In the Kibera project the municipal administration even took over the role of managing the latrine emptying service (from the community), although this intervention resulted in the eventual collapse of the service.

The role of the municipal council in project implementation was more critical - including responsibility for overseeing project implementation, planning with the community, provision of permits for water connections, access to excreta disposal facilities (sewerage system) and support for solid waste collection.

In both cases this role was (or is still) effectively carried out during project implementation but after project completion, for example in Maina Village this support has not been regular. Solid waste collection and sewer maintenance and enforcement of the public health act has not been consistent. The environmental sanitation in Maina has therefore deteriorated noticeably.

Lessons Learned

(a) Community participation in urban informal settlements is being complicated by various factors. Tenant communities often lack the initiative to improve their living environments. This is a result of the heterogeneity of urban communities and insecurity of tenure of both landlords and tenants. In the case of Maina village, several factors led to the collapse of the community initiatives. Key factors were the change in tenure and enlargement of the settlement area.

(b) NGOs can play the role of catalyst and be able to guide communities. The chief catalyst role of an NGO is to create self-reliance among the people they are working with. The role of NGOs in building community commitment to improving environmental sanitation in informal settlements has proved to be critical, however sustainability of efforts relies largely on continued intervention of internal (CBOs) or external (NGOs) support groups. In Kibera, the continued presence of KWAHO has availed the community continued support. A clear example is KWAHO´s agreement to revive the latrine emptying service.

(c) Technology selection should be based on a careful analysis of sustainability in terms of technical, social, economic and institutional factors. In the case of the latrine emptying machine, it was technically suitable for Kibera's narrow paths and difficult to access latrines. However, it failed to function because of inadequate institutional arrangements. Lack of adequate institutional arrange ments and unclear ownership led to the local administration taking over the management of the service. This eventually resulted in the collapse of the service.

(d) Security of tenure is a critical factor in upgrading or improving infrastructure in informal settlements. If people are expected to participate in the improvement of their living environment it is particularly essential. In any project involving the installation of physical infrastructures (often with community resource contributions), the permanency of such infrastructures should be guaranteed against unnecessary alterations. This guarantee can be forthcoming only if there is a well defined land tenure system and a mutually agreed and legal physical plan (7). In the case of Maina village, the project implementation agreement between Nyahururu municipal council and DANIDA was not binding, and several major alterations have since been recommended and undertaken. Although the community has successfully argued against major changes that would affect the investments made under the project it should be noted that is important that project activities are tied up to long term development plans.

(e) Training and health education are important elements of environmental sanitation projects. Training in operation, maintenance and overall project management is an important component of any community based project. Training needs assessments must be carried out as part of project design. If training is conducted as a continuous process, involving the community and all other parties it can ensure sustainability and effective use of facilities. In Maina and Kibera training has been carried out by KWAHO, but the results indicate that health education training was inadequate. Residents of Maina have not adequately understood the use of the sewer system and have therefore contributed to its malfunctioning. Two examples are the continued use of inappropriate materials such as maize cobs in flush toilets and non-payment of water bills.

(f) Adequate implementation arrangements should be made by donors, project managers and NGOs, if sustainability of project interventions is to be ensured. In the case of Maina village, the municipal council has not continued to provide a solid waste collection service, or to maintain the sewers. This has resulted in the community abandoning clean up days, and solid waste is now dumped in any convenient location. The inadequate maintenance of the sewer has also discouraged residents from paying for water bills and using the system properly.

References

1. Nairobi's Informal Settlements: An Inventory. USAID/REDSO/ESA, March 1993, page 5
2. Ibid
3. NGO Support to Informal Settlements. A Case Study of Kibera, Nairobi, 1991, page 10.
4. Ibid
5. Daily Nation, Thursday June 23, 1994.
6. NGO support to Informal Settlements. A case study of Kibera, Nairobi, 1991
7. Community Mobilization in Sanitation Projects. A Case Study of Maina Village, 1994, page 17.

COMMENTARY ON *THEME I.*

Ms. Almedon remarked that she was worried that women were under-represented at the conference until she heard *Ms. Kunguru's* paper. She asked whether NGO's are willing to address local concerns in their efforts to improve sanitation in periurban areas. *Ms. Kunguru* replied that NGOs should be able to look for appropriate local designs and not arrive with preconceived notions.

Mr Walakhwa of Eldoret Municipal Council voiced approval of local participation. Eldoret has lots of programmes to promote planning awareness in the periurban communities. But he has seen the danger of relying too much on local people; they tend to express needs and then loose interest. *Ms. Kunguru* replied that there is a great need to discover how and for what purpose local communities are formed.

An unidentified participant said that he had often seen NGO's come into periurban areas. Even if they have unlimited resources, what is their purpose? *Ms. Kunguru* replied that it is not an issue of money but one of commitment. Communities can make their own contributions in kind and in spirit. The best NGO's can do is bring awareness, information and ideas. When given ideas communities can generate their own resources. A 50/50 relationship with an NGO gives a community a greater sense of ownership and autonomy.

Mr Haraka of Nakuru Municipal Council related his own experience as a local official and the role of NGOs in periurban communities. He explained that local officials appointed by NGOs often abscond because they are so hard working, and cannot make time to carry out imposed requirements. He told an anecdote of an NGO-sponsored visit to a Tanzanian village the main product of which was the conclusion that water can be hauled by donkeys. He concluded by emphasizing the importance of zoning regulations in Nakuru and other urban areas.

THEME II : SAFE WATER ENVIRONMENTS RESEARCH IN ELDORET

Chairman: Professor H N K arap Mengech

Professor Mengech extended a warm welcome to guests from outside Kenya and wished them a pleasant stay. All were bid welcome to Eldoret which is not only the venue for the conference but also for the Safe Water Environments Research Programme that was conducted between October 1993 and April 1995.

This research was conducted by staff from the university of Linköping in collaboration with the faculty of Health Science and Department of Anthropology, School of Social, Cultural and Development Studies, Moi university. The research programme at Moi university involved six anthropology students and four medical students under the supervision of their academic members of staff.

The chairman hoped that what will be discussed in the conference, especially concerning water and living environments of the less fortunate parts of our urban centres, will lead to the formulation of recommendations which can assist implementors and consumers and provide a direction for future research.

Paper 4. SAFE WATER ENVIRONMENT IN ELDORET TOWN: THE SOCIAL DIMENSION

Joshua J. Akong'a and Harrison M.K. Maithya
Department of Anthropology, Moi University, P.O. Box 4606, Eldoret, Kenya

Introduction

The problem of availability of water in desirable quantities and quality has become a major problem especially in periurban areas in technologically less developed countries like Kenya. In Kenya for example, 30 to 40% of urban residents are found in slums where there are no adequate services such as the provision of piped water, drainage, street lighting, pavements, roads, garbage collection, public toilets and regular police patrols, among others. The poor sanitary conditions prevalent in periurban areas can partly be accounted for by the inadequate services extended into these areas by local councils and the wide-spread poverty found among urban residents. In Kenya, it has been estimated that 60% of urban residents are poor, making it difficult for them to pay for services available in other parts of the urban. Many urban residents especially in periurban slums are unemployed, underemployed and generally low income earners. A study carried out in 5 slums of Eldoret town for example, revealed that 21% of the respondents were unemployed and yet 74% are married, 73% with at least 1 to 4 children and 63% with 1 to 4 other dependants besides their children (Wanyonyi, 1994:55). It is, therefore, not surprising that small rooms found in slum areas of urban centres in Kenya are over crowded, and the environment generally depressed and unimpressive.

These are conditions that become breeding grounds for what Oscar Lewis would term the culture of poverty as manifested in American towns but which we could refer to as poverty of culture in technologically less developed countries such as Kenya. A situation of poverty of culture arises when people are not sure what is good and what is not due to an absence of a generally accepted standard of judgement or due to a feeling of helplessness arising from the lack of the means for doing something about what they may recognize as not acceptable. The very low standard of personal and environmental hygiene observed in slums of Kenya is therefore something slum residents can alleviate given some initiative. The provision of wells/ bore holes on each plot and the spontaneous and haphazard settlement in slum areas are an aspect of improvisation of facilities where the formal sector has been unable to cope with the demand for low cost services.

Objectives of the Study

There were two dimensions. The social dimension consisted of the assessment of hygiene behaviour that might lead to contamination of water and the environment. The microbiological dimension concerned the assessment of contamination routes including groundwater contamination and measurements of microbiological organisms available in water for domestic use in the study area.

This paper will restrict itself to the social dimension of Safe Water Environments in Eldoret Municipality. This aspect of the study was carried out by six anthropology students and four medical students.

Methodology

The study was carried out in two periurban slums, Langas and Huruma (Mwenderi) in Eldoret town and a village named Kimumu. Both Langas and Huruma have many characteristics in common including water sources, sanitation facilities, housing conditions and also economic characteristics of the population. Population density in these areas is high and the majority are in the low income category.

Non probability sampling procedures were employed to select 15 compounds in both Huruma and Langas that were included in the study. The compounds were selected on the basis of household population and availability of water sources, mainly a well. These compounds all had semi-permanent houses which have mud walls, cemented floors and corrugated iron sheet roofing; except for one house in Langas and one in Huruma. In each compound the houses are constructed in rows of two or three facing each other.

The population ranges from 8 to 40 people per compound. Most of the households live in single rooms which are fairly small and congested. Most of the tenants earn their living as casual labourers, hawkers, shop attendants, artisans or by selling illicit brew (Chang'aa and Busaa). Thus their source of income is highly unreliable. In some cases, the landlords or landladies live in the compounds. Most of the facilities in the compound such as pit latrines, bathrooms and water sources are communal. Most compounds have one or two pit latrines and a bathroom all of which are poorly maintained and poorly constructed semi-permanent structures (wooden wall with or without roofs).

The main source of water is the well and some plots have two wells. The well is usually located in the middle or at the edge of the compound. Some wells have wooden covers, while others are open. The surrounding area of some wells is either raised or cemented but none of the wells were lined. Domestic animals which including cows, goats, pigs, dogs, ducks and chicken are also found in most compounds. It is therefore not uncommon to find animal droppings and human faecal matter littering parts of the compounds. The socio-economic characteristics, environmental, housing and sanitary conditions are thus typical of low income groups and conditions under which the urban poor live.

In order to obtain data on the social dimension of the study, complementary techniques were employed so as to obtain both qualitative and quantitative data. These were observations, focused group discussions, informal and formal interviews; the latter through the use of a standard questionnaire. Structured and unstructured observations were made at three levels viz: at individual/household level, interaction between individuals in the compound and the compound in general. Structured observation focused on the activities at the water source, water en-route or transportation of water from source to the house and water related behaviour in the house. Boot and Cairncross (1993) have observed that structured observation is very useful in collecting water and sanitation related data where there are few households.

Of specific interest was tenants behaviour around the water source particularly wells. These included activities such as drawing of water, where the container for drawing water is placed, precautions taken, the area surrounding the well and other activities such as washing of cloths, utensils and children playing around the well. Observations on how water is transported from the well to the house and the containers used was of much interest to the study. Observations were also made on water storage in the home and other water related behaviour in the house. Latrine utilization and related behaviour was also a major area of our focus.

Spot checks and odd hour observations such as early morning, late evening, overnight as well as at the weekend were later employed to capture unusual behaviour. Observation was used in both dry and wet season and throughout the field exercise. Informal interviews and focused group discussions carried out after the initial stages of observations yielded data on knowledge, attitudes and other aspects of hygienic behaviour.

Formal interviews with willing household members in each compound were conducted after a prolonged period of observations. The individual interviews were divided into two or three sessions in order to maintain the interviewee attention. The questionnaire instrument was pretested and questions revised. Also the content of interview questions evolved after the long period of field observations and experience. This approach has been found to be very useful in obtaining in-depth information from informants (Drangert, 1993). Informants knowledge, perceptions about water quality, management of their environments and sanitary facilities were some of the areas of interest. Scenario questions provided insights in assessing water quality and also sanitation related behaviour. Water sampling for microbiological measurements through laboratory analysis was also employed to complement observations. This method was particularly useful not only in assessing contamination routes but also validating observations made on tenants behaviours that may result in water contamination.

The methods employed as stated above enabled the research assistants to collect both qualitative and quantitative data. Apart from some field observations, perusing through field notes, discussions papers and reports compiled by the research assistants, selected questions on the standard questionnaire were analyzed to provide some statistical infor-mation. The total number of questionnaires available for analysis were 65. However, the 12 questionnaires administered at Kimumu which is also part of Eldoret municipality, and which was supposed to be used as a control group representing a rural situation, were not analyzed. This was because the authors of this paper did not consider Kimumu to be a typical village. It is a settlement area where residents practise agriculture but use well water instead of spring or river water sources as is the case in a typical village.

This study therefore, represents what is possible in terms of collaborative research between social and biological sciences on health related issues.

Results

Some of the findings of this study tally with what Wanyonyi (1994:92) found in her study. When she asked her respondents what they thought were their main problems, the following responses were obtained: 67% said water, 49% said security, 48% said poor sanitation, 32% poor roads. Other problems included lack of electricity, congestion, noise, drunkenness, and air pollution. In his study of housing, water and health in Kisumu municipality, Kasoa (1989) found similar conditions.

Perception of a clean compound can vary between individuals. When asked what constitutes a clean compound our respondents gave the following responses:- 64% (34) said it is without litter, dust, dung and/or faeces, 21% (11) said the compound should have trimmed grass, 21% (11) said it should have a good toilet/latrine which is not visible, 13% (7) mentioned good drainage. Others mentioned rubbish pit/dust bin 11 (6), piped water (5), house walls should be well constructed (4). Only 1 respondent said that the water source should be protected.

Although there were few responses to the question, the generative nature of these responses provide an indication of the attitudes residents/tenants have towards the physical environment in which they reside. Ranked high is the physical appearance of the compound, in terms of cleanliness and trimmed grass, followed by functioning utilities such as latrines, drainage, rubbish pits and so on. The implication is that these things are generally lacking or are not provided in adequate quantity and quality and it suggests that residents of slums reside in these areas as they have no alternative given their socio-economic circumstances.

Water related behaviour. If we want to understand water related behaviour and its implication on health, peoples assessment of water quality is a good starting point. How people determined whether water was good for drinking is presented in table 1.

Table 1: Assessment of water quality

Response	No. of answers
Observation*	43 (81%)
No smell	8 (15%)
Hard to tell	6 (11%)
All flowing water is clean	1 (-)
When not tap water	1 (-)
Depends on source	7 (13%)
Taste	12 (23%)

** observation refers to what the consumer can see (i.e. particulate matter, colour, turbidity, presence of organisms). The implication is that water that is colourless or clear must be clean and therefore safe to drink*

These responses confirm field observations and informal interviews which revealed that during the dry season tenants would mostly draw water early in the morning and late in the evening when the turbidity is low. A scenario question on how to assess water quality if offered in a strange environment, for example in a hotel yielded the following results: 21% (11) of the informants said they would observe, 26% (14) said they would use taste, 9% (5) said they would assume the water is safe while 2 informants said they would know by smell. Others said they trust hotels, or ask if it is tap water. One informant said that Legio Maria (a religious sect) carry boiled water.

From the above responses it can be concluded that the use of the senses of sight, taste and smell is paramount in gauging the quality of water. Appearance, taste and smell are popular assessments of water quality. In his study of water in Sukuma land, Drangert, (1993:108) found similar elements very important in assessing water quality. Households would use some or all of these factors to assess their water. Similarly Mukherjee (1990) found the same perceptions about good water. On the basis of a countrywide study in India, she concluded that water that is visually clean, tastes sweet, has no odour, and cooks food well and quickly is defined as 'good drinking water' (cited in Boot and Cairncross, 1993). This boils down to some stereotypical assumptions that "clean" water is safe; tap water is clean and safe, and all these are based on trust. This points to a dangerous lack of knowledge of potential infection inherent in water; it is easy to transmit infectious water borne diseases having such assumptions. This is more so when it is noted that the majority of slum dwellers in Eldoret Municipality use wells located in the compounds of their residence as sources of water for domestic use.

In a study carried out by German Environmental Consultants in Eldoret (1989:4 Vol. 1 & 1989:19 Vol II) it was found that 59% of people use dug wells while only 36% especially in Langas use Municipal Council water kiosks for drinking and cooking requirements throughout the year. 4% use streams throughout the year as the only source of water. The study carried out by Wanyonyi in Langas, Mwenderi, Kamukunji, Hill school village and Ngomongo revealed similar results:

Table 2: Source of Water

Area	Municipal Tap	Dug Wells	Other
Langas	42 %	55 %	0 %
Mwenderi	100 %	0 %	0 %
Kamukunji	46 %	29 %	25 %
Hill School	0 %	100 %	0 %
Ngomongo	8 %	69 %	22 %
Total	38 %	52 %	8 %

Source: Wanyonyi 1994:77

Regardless of the source water can be contaminated. Esrey et al (1990) note that contamination of water may occur in different ways and at different stages. For example, the quality of water may be high at the source but low in the household (cited in Drangert, 1993:121). When the informants were asked ways in which water can be contaminated their responses tended to point to the source. Few tenants seemed to be aware of water contamination when its being transported or stored in the house. Asked whether there are some people who try to contaminate water, results in Table 3 were obtained:-

Table 3: Are there some people who try to contaminate water?

Response	Number	%
Yes	31	58
No	19	36
N/A	3	6
Total	53	100

Data in Table 3 suggest that tenants are aware some people may contaminate water. Water was said to be contaminated in several ways which include: using a dirty container to draw water, spillages, storm water, via wind blow debris, by children, by drunkards, by tenants washing clothes close to a well, because the well is uncovered and through carelessness. Few people seemed to be aware of underground contamination from pit latrines. Phage analysis indicated that pit latrines may be a source of well water contamination (see paper 15 in these proceedings). Although informants are aware of sources of water contamination, they do very little to prevent it. For example, tenants were occasionally observed washing clothes and pouring the waste water not far from the well. Drawing containers would also be placed carelessly on the wet ground. Occasionally children would play in the pools near the well particularly during the wet season.

However, one may not have control over other people´s behaviour. Asked whether anyone can be excluded from the well about 40% (21) of the informants said one can be excluded while 57% (30) said one cannot be excluded from the well. Those who responded in the affirmative gave a number of reasons including the fact that someone could poison the well. These are some of the efforts to maintain water quality. Others said that when a well is being cleaned during dry season, when water level is low, non-residents could also be excluded.

Water is a common resource whose access has social implications. Village informants reported that water is a natural resource, a gift from God and one can therefore not deny others this resource. In the periurban areas those who said one could not be excluded from the well explained that all tenants are entitled to use the well because water is included in the rent. Others said that they cannot refuse anyone water, one can only be refused if they are not a tenant and neighbours have to pay before drawing water. If there is no caretaker it may also be very difficult to exclude people from the well. These responses point to the helpless situation slum dwellers find themselves in. The attitude of the tenants is that well protection is the responsibility of plot owners. 85% of the tenants felt that it was the landlords responsibility to clean the well. Observations showed most of the wells were not cleaned for more than a year. Thus even if the slum dwellers wished to improve the quality of well water at the source they lack power both socially and economically.

Some of the tenants, however, make efforts to mobilize others in the compound to protect wells. One tenant in Langas organized the covering of a well. He brought timber and paid for labour as he felt that too much surface run off was getting into the well. Another tenant organized the digging of a rubbish pit and this encouraged other tenants to throw their solid waste in it.

Asked whether anyone can be excluded from getting near the well, 77% (41) said yes while 17% (9) said no. Of those who said yes, 74% (39) talked of children not being allowed to play close to the well, whilst drunkards are excluded. Others indicated that since the well is not locked, and compounds are not fenced, it is difficult to prevent people getting close to the well, after all there are no regulations. The foregoing responses imply a distinction between the use of a well and exclusion from getting near it. However observations revealed a different situation. It is also important to note that informants did not mention keeping animals away from the water source.

Although most tenants use well water for all domestic purposes, this may not be out of choice but partly as a result of inadequate or non availability of other water sources. 53% (28) of the informants appreciated that tap water was better and said it was worthwhile regardless of distance, time and cost. Only 13% (7) of the respondents said they did not think it was worthwhile to go for tap water. Those who preferred tap water despite problems and cost said among other things that well water is not safe for drinking as children throw objects into it. Others said that well water is salty, has a bad smell and taste, and the well has no cover and is near a latrine. Tap water was also said to treat stomach ache and is treated with chlorine. Despite the use of tap water involves extra changes, tenants are willing to pay for the extra cost. A 20 litre container cost between 50 cents and 80 cents. It can be concluded from the above responses that inadequate or non availability of tap water is one of the major limiting factors to its use.

In many communities, water drawing and water use may be regulated by certain cultural factors. For instance in rural areas, it is women who cook, draw water from wells and transport it on their backs or heads and use it for cooking and washing. Men may occasionally draw water but transport it by their shoulders, wheelbarrows, carts or bicycles. This may, however, not be the case in urban areas. Informants were asked whether there are customs about how and who draws water.

Table 4: Customs related to water drawing

Responses	No. of Informants	Percentage
Yes	17	32
No	33	62
Don't know	1	2
N/A	2	4
Total	53	100

When a rural society becomes urban there are many associated cultural changes (Mathenge et al, 1987:59). The respondents who said there were no customs related to drawing water, often qualified this by saying that there were customs of rural origin but these did not apply in the urban setting. Respondents who said customs existed identified the factors such as men in African culture are not supposed to fetch water or men cannot go where women bathe. Others felt that women should be well dressed at the well. Other informants said that a person cannot draw water while mourning a spouse and women who are menstruating or have recently given birth should not draw water. Others felt that water for drinking and bathing should be drawn from different sources. Field observations revealed that it was mostly women and young children who draw water. These customs may have some bearing on water quality.

On the relationship between water and the mosquito 79% (42) of respondents said there is a link, 6% (3) said there is no relationship while 11% (6) said they did not know. The majority recognize the danger of stagnant water and its association with mosquito breeding especially during the rainy season.

Water and sanitation related diseases include various types of diarrhoea, and malaria among others (Boot and Cairncross, 1993). On the other hand domestic and environmental hygiene relates to household cleanliness, garbage disposal and animal management (ibid). Although the informants are aware of dangers of stagnant water and generally of an unclean and waterlogged environment, this largely thought of as a wet season problem. The compounds are usually dirty with pools of stagnant water due to poor or clogged surface drainage. Tenants do not find time to clean their compound or do not consider this their responsibility, or simply neglect their compounds. This is exemplified by responses to a question on what happens to the well when it rains. 13 of them said surface water can enter, 21 informants said nothing would happen, whilst others said the pit latrine is at safe distance away so there will be no problem. The few responses to this question indicate the extent to which few people are willing to give any thought to the water they consume; partly because there is little they can do about it and also because the alternatives are distant to them. During the wet season, storm water gets into some of the wells as their head walls are not raised above ground level and even when the well is covered this does not guarantee protection.

Cooperation with other tenants to improve compounds may be difficult as some people are not willing to make a communal effort or lack the resources required. Some tenants do not consider themselves permanent in these compounds and may therefore opt to leave the compound when conditions become unbearable. Others may stay on as the conditions in the area are just about the same. Tenants are, however, aware of ways to improve water quality. Informants mentioned boiling water, decanting, filtration and chlorination. One way of decanting reported in the village is putting "erut" leaf in stored water so that dirt can settle at the bottom. [This species was not identified but could probably be a member of the *Moringa* family. Eds] Some tenants said that they boil water only during the rainy season when water is perceived as dirty. Observations showed that few tenants boil water for drinking. Decanting of water in the house was done by some tenants and they said that it makes the water clear. Thus as we have already seen observation becomes important in judging water quality.

Sanitation related behaviour. Behaviours and facilities associated with the safe disposal of human excreta and use of water for hygiene are important measures for eliminating certain diseases. However, in most urban development programmes although water supplies are constantly upgraded, sanitation has not re-ceived the same attention (Arnold, 1978). This situation exists in Langas and Huruma (Mwen-deri) as the municipal council has made efforts to supply the residents with tap water via "water kiosks" but not much attention has been paid to sanitation. For instance, German Environ-mental Consultants (1989:83 Vol I.) revealed a major problem in the disposal of faeces. Only 0.5% of slum residents have flush toilets, 2% have no toilets of their own and 84% use communal toilets, while some 13% use the bush or a neighbouring compound. The hygienic conditions of the available facilities leave a lot to be desired because they are not regularly cleaned (ibid, 4).

In our survey, the respondents were asked what should be done if they found faeces on the latrine floor. 47% (25) said that they would do nothing, 36% (19) said they would clean it up if it was from a child, 4 of the informants said they would not use such a latrine, while others said that they cannot clean other peoples dirt or else they would wear shoes to use the latrine. This is exemplified in statements such as "Nikipata mavi juu ya choo nitatafuta mahali nijisaidie ikiwa ni haja kubwa ama niende nikajisaidie mahali pengine". ["If there was faeces on the latrine slab I would find another latrine to use, or if I needed to urinate I would go else-where." One implication being that you can urinate elsewhere than a latrine. Eds]. These responses show that only a third of the respon-dents (36%) would do something about the unsanitary conditions. The other respondents avoided the question altogether or would do nothing about the unsanitary conditions.

In view of this the respondents, the majority of whom were tenants, were asked if there was anyone responsible for cleaning the latrine. The following responses were recorded:

Table 5: Is there anyone responsible for latrine cleaning/ sweeping?

Response	No.	Percentage
Yes	29	55
No	23	44
N/A	1	1
Total	53	100

From the table above it is obvious that the cleaning of the latrines is not taken seriously by the house owners. In cases where there is a specific person responsible, it is either the landlord/landlady or a close relative. The land-lords probably do not wish to hire anyone to keep the compound clean. It is also the case that tenants do not consider the cleaning of the latrines as their responsibility but a respon-sibility of landlords as they consider the cost of these services included in the rent. The end result of these attitudes is the observed filthy environment and pit latrines particularly during the wet season.

On how often the latrine is cleaned, few responses were obtained: 7 informants said two to three times a week, 17 informants said once a week, 11 said once in a while, 2 said once in two weeks while 1 informant said when it is dirty and another one said whenever it is dirty. Responses to this question confirm the observation we have already made about the general lack of environ-mental hygiene in the slums of Eldoret.

Some of the consequences of poor cleanliness of latrines include:

- Flies	63 %
- Smell	73 %
- Flooding	23 %
- Problems with children using latrines	56 %
- Emptying	16 %
- No problem	1 %

Source: German Environmental Consultants, Vol. I, 1989:83.

Arnold (1978) observes that pit latrines can break the cycle of infection but do not succeed when they are not consistently used which can be the case when the physical structures are unattractive to the users. In Eldoret periurban slums latrines commonly have semi permanent wooden super structures. These may in turn harbour insects that lead to serious health problems. Studies show that flies play a very important role in the transmission of a number of diseases among them conjunctivitis (eye diseases) (Boot and Cairncross, 1993). Their eradication is there-fore necessary if we expect to reduce disease. Asked what should be done about flies found in the latrine respondents gave the following information: 53% (28) said they should be sprayed, 19% (10) said latrine should be kept clean, 4 informants said they do not know, 5 said they did not think it is a problem. Other responses were given which include use of soda ash or ordinary ash, covering the squat hole, moving house, seeking the landlords help, and doing nothing. Responding to the next question on whether they had tried any of the solutions suggested, 60% (32) said they had not tried, 15% (8) said they had. These responses imply that although respondents know ways through which the problem of flies could be solved, they either do not consider flies as a serious health problem or they are unwilling to take any action as they do not consider themselves permanent in these compounds. Failure to take action may be due to lack of resource, for example, cash to buy sprays or simply because it is not perceived as their individual or household problem since latrines are usually shared with other families.

Children's faeces. Where faeces are left, be it adults' faeces in the bush or children's faeces left in the courtyard, it may increase incidence of certain diseases that result in morbidity and mortality (Rahman, et al, 1985 cited in Boot and Cairncross, 1993).

Table 6: Whether child faeces is left to the dogs

Responses	No	Percentage
Yes	9	17
No	39	74
Don't know	1	2
Have not seen it	2	4
It is left uncleaned	2	4
Total	53	100

From the above table, it is clear that people would not wish to have children's faeces picked up by dogs, at least not intentionally. However, because many children's faeces are not usually cleaned away, dogs end up picking them up. Some respondents mentioned that it depends on the age of the child. Faeces of big children must be cleaned away but those of small children may be ignored. Residents who have some space or garden nearby direct their small children there, out of the belief that faeces are harmless and should not be mixed with those of adults most of whom are not even family members. Observations showed that some mothers were not quick at collecting children's faeces while others were prompt.

As to whether it is safe to throw children's faeces into the latrine 11% (6) of the respondents said no, while 83% (44) said yes. Those who were positive gave reasons for cleaning up: 13 of them said it minimises diseases/infection, 12 said it makes the compound clean, 3 said it minimizes bad smell while another 3 said small children can eat it. Those who thought it was not proper to throw children's faeces into the latrine explained that children stools should not be mixed with those of adults, child stool has to be hidden because there is danger of a witch picking on the stool of a particular child and since everyone sees the other go to the latrine, faeces left in shallow latrines can be picked by people with ill will. Some said that they throw children's faeces into the drain. Those who have no bush that can provide the children with some cover therefore think that of all the ways of faeces disposal, it is better in the latrines rather than the drain. However, on several occasions mothers were seen throwing children's faeces somewhere in the compound or in the open fields.

When respondents were asked what they would do if a child has defecated on the latrine slab the following information was generated:- 23% (12) said they would tell the mother of child, 43% (23) said they would clean it up 36% (19) would use the pit latrine and leave it as found, and 8% (4) would punish or warn the child.

Concerning potential infection, respondents were asked if there is a difference between a normal stool and a watery stool. 87% (46) of the respondents said yes while only 2 said no. Asked if they take any special precaution when the stool is watery, 79% (42) said yes, while 3 said no. Of those who said yes several alternative precautions tried were mentioned as in table 7:

Table 7: Precautions taken when stool is watery (N=53)

Responses	No	Percentage
Give child hard food	1	-
Try medicine/sugar, salt or wheat flour in water	10	19
Take to hospital	22	42
Give a lot of water	1	-
Wash child	2	-

The poor response to this question is worrisome because most people do differentiate between the type of stools, but few do anything about the danger they perceive.

Water contamination can also result from improper handling of soiled clothes. When asked where clothes are washed 42% (22) of the informants said next to the well, 13% (7) said at the door step, 23 % (12) said far from water, 11% (6) said near drainage, while others said inside the house, at a common washing place or at river Sosiani.

From these responses, it can be surmised that there is an ever present risk of contaminating well water given the clogged drains, and poor practices regarding washing soiled clothes and cleaning up children's faeces. Water poured in the compound is likely to drain to the well. In fact much of the waste water is not poured far from the mouth of the well. These are risk factors for the transmission of water-borne diseases such as diarrhoea and hookworm infection (Ouma, 1995; Wanyonyi, 1994:81). The situation is rather critical considering the fact that wells are on average not far from pit latrines and that the wells are not properly covered and protected. The problem is even more critical during the wet season.

According to the German Environmental Consultants (Vol I, 1989:27) many well covers are missing, wells are inadequately lined, linings are not extended above ground level or plastered, well aprons and drainage is inadequate and water is drawn by bucket and rope which introduces pollution to the well. Observation confirmed this situation.

Due to these poor hygienic conditions, the potential for morbidity is an ever present reality. However, according to the Germany Environmental Consultants (Vol. I, 1989:92) morbidity in Eldoret slums is lower than expected. This was said to be due to a combination of high altitude, local temperature, seasonality of rainfall and the skeletal municipal services provided. It could also be due to immunity developed by the slum dwellers over time.

When the respondents were asked how often children suffered from diarrhoea, the following responses were recorded:-

Table 8: How often do children suffer diarrhoea? (N=53)

Responses	No	Percentage
Rarely	15	29
Frequently	14	26
Weekly	4	8
In two weeks	1	-
Not seen	2	-
Don't know	5	9

The reasons given for the frequent diarrhoea problem in children included unsafe water (28%), poorly cooked food (30%), unsanitary conditions (17%), teething prob-lems (17%), eating dirty things/soil (13%). Other reasons include flies and dirty utensils, adultery, having no fixed eating time, over feeding, worms, and that when using water next to a latrine diarrhoea comes by itself. These responses show that tenants do not consider an unhygienic environment as the only cause of diarrhoea. There are other causes, among them cultural ones which may not be perceived treatable in hospital. This may again explain the low morbidity when only health centre records are used.

The respondents, however, felt that there was a way by which this problem could be prevented. The prevention measures included improving environmental sanitation, using piped or clean or boiled water, eating well cooked food, teaching parents better child care through personal hygiene and by resorting to a doctor/hospital or using traditional medicine. Other prevention measures include taking care of children so that they do not eat dirt, storing food safely, being moral and offering material assistance to the poor.

Conclusion and Recommendations

This study concurs with the findings of the German Environmental Consultants (Vol. I, 1989:8) in that slum residents in Eldoret (and everywhere in Kenya) need increased public health awareness. Environmental conditions in the study area are conducive to water contamination both on the surface and underground. The behaviour patterns and attitudes of the slum residents influences and predisposes them to potential infection because the standards of waste and faeces management are very low. It was also observed that the implications of infection are not well appreciated. Adequate sanitation is, however, a pre-requisite for the preven-tion of environmental pollution as well as water related diseases. If there is any role the Local Authority Development Programme operating in Kenyan municipal councils can play, it is therefore to work with urban residents in sensitizing them on the dangers they are exposed to so that even though the local authorities have limited resources to provide basic services, the residents can help themselves through personal and environmental hygiene in the compounds they reside. It has also been demonstrated that the multi-dimensional nature of safe water and other related health issues in periurban areas can only be understood through collaborative interdisciplinary research such as this study, where anthropologists can work hand in hand with medical practitioners, microbiologists, policy makers and implementors such as local authorities and with the residents themselves (Akong'a, 1991). The implication is that action to alleviate the problems and implement the solutions identified cannot be achieved by one sector but by all those concerned in collaboration.

Acknowledgement

This paper is based on the field work of six anthropology students undergoing their final degree programme at Moi university. These were Jerry Okal, Maurice Kong'ong'o, Bernard Mundia, Violette Naanyu, Malei Nthiwa and Anthony Macharia, under the field supervision of Mr. Harrision Maithya.

References

Akong'a, J. 1991 Participatory Action Research in the Development Process, in Orieko Chitere and Roberta Mutiso (eds) *Working With Rural Communities.* Nairobi University Press.

Arnold, P. (ed). 1978. *Sanitation in Developing Countries.* Chichester: John Wiley and Sons Ltd.

Boot, M.T and S. Cairncross (eds) 1993. *Actions Speak: The Study of Hygiene Behaviour in Water and Sanitary Projects*, London: IRC International Water and Sanitation Centre and London School of Hygiene and Tropical Medicine.

Broome, J.M. et al. (eds). 1987. *Sanitation Field Manual for Kenya*, Nairobi: Government of Kenya, UNDP and World Bank.

Drangert, J-O 1993. *Who Cares About Water? Household Water Development in Sukumaland, Tanzania.* Linköping, Linköping Studies in Arts and Science.

German Environmental Consultants. 1989. *Eldoret Municipal Council Sanitation Study*- Vol. 1, Report Annex A September, 1989, and Vol. II Annex B. Study on Socio-Economy and Public Health Pre-Feasibility Report.

Kasoa, M. 1989. *Political and Administrative Factors Influencing Kisumu Municipal Council in Service Provision - Housing, Water and Health*, M.A Thesis, University of Nairobi.

Nyaoro, W. 1992. *Water Supply and Consumption in Eldoret.* M.Phil. Thesis, School of Environmental Studies, Moi University.

Ouma, W. 1995. *Factors that Influence Prevalence of Hookworm Infection in Seme, Kisumu District.* M.A. Thesis, University of Nairobi.

Wanyonyi, R.N. 1994. *The Urban Poor and the Environment in Eldoret, Kenya.* M.Phil. Thesis, School of Environmental Studies, Moi University.

Paper 5. A LONGITUDINAL STUDY OF THE RELATIONSHIP BETWEEN WATER QUALITY, HYGIENE BEHAVIOUR AND CHILDHOOD DIARRHOEA IN LANGAS - ELDORET, KENYA

K. K. Kamar, Department of Microbiology and Parasitology
Faculty of Health Sciences, Moi University, P.O. Box 4606, Eldoret

Introduction

The prevalent environmental conditions in Langas and in many other periurban areas in the developing countries ease the transmission of diarrhoeal disease. Faecal contamination of drinking water is probably a major route of spread of these diseases.

Much of the developing world is plagued with a high level of faecally contaminated drinking water, endemic diarrhoeal disease and an acute shortage of resources for public health services (Shiffman et al., 1979). Yet Huttly (1990) observed that the simple provision of a clean adequate water supply and a good sanitary facility will not automatically result in significant improvement in health. Notwithstanding the unpredictability of the result of many water-sanitation studies, the general picture suggests that access to water and improvements in hygiene may have a greater impact on good health than improved water and sanitation facilities.

Briscoe (1984) described in a stimulating theoretical paper the basic conceptual difficulties in performing studies involving multiple transmission routes. In his paper, volunteers in a hypothetical experiment ingested various doses of an imaginary diarrhoea-causing pathogen with a log linear dose-response relationship and showed that reductions in the ingested dose do not necessarily lead to proportionate reduction in the incidence of disease. He concluded that the elimination of one of several important transmission routes cannot be expected to cause a measurable reduction incidence of disease.

Similar suggestions have been made by previous work notably that of Shuval et al (1981) who argued that the population's "response" to a "dose" of water and sanitation depended greatly upon the general levels of faecal contamination, hygiene and socio-economic development.

Bradley's (1977) classification focuses on how interventions in water and sanitation may affect health and suggests that increasing the quality of water available to a household can reduce the water washed diseases only if more water is used for handwashing.

Studies by Feachem (1984) and Caincross (1987,1989,1992) show that most significant impacts on disease stem from changes in hygienic practices such as handwashing, reheating left over food; improved water storage, disposal of child faeces, and washing and handling of utensils rather than merely from the provision of clean water. The impact of handwashing upon diarrhoeal diseases has been particularly well documented. Han and Hlaing (1989) reported a 30% reduction of diarrhoeal diseases among children under five years due to increased handwashing. Khan (1992) reported a 69% reduction in secondary infection of Shigella because of increased handwashing with soap; washing without soap produced far less impact. WHO (1993) studies of hygiene improvements linked to water and sanitation use have shown a reduction in diarrhoea incidence of between 14-18%.

A study in rural Bangladesh (ICDDR 1991) concluded that three key behaviours could lead to the greatest reduction in the incidence of diarrhoea namely; disposal of faeces, handwashing, and safe water handling to prevent faecal contamination. In a study in Dhaka, Bangladesh, Clemens and Stanton (1987) noted that acute diarrhoea imposes a high burden of morbidity and mortality upon children in the developing world and that the relationship between specific water and sanitation behaviours and the risk of diarrhoea rests largely on anecdotal evidence, providing little guidance for the rational formulation of educational interventions. They reasoned that an educational intervention to improve practices related to water and sanitation would be most likely to succeed if it consisted of relatively few messages whose prescribed behaviour occurred in the community with empirical correlation to reduction in childhood diarrhoea.

The present study was aimed at testing the influences of the bacteriological quality of water, sanitation, and the mothers hygienic behaviour on childhood diarrhoea in children aged between one and 12 years residing in Langas. The study had the following objectives:

◊ To determine the bacteriological quality (BQW) of well water in Langas.

◊ To investigate the relationship between BQW and the distance to latrines.

◊ To relate the BQW to the incidence of childhood diarrhoea.

◊ To determine the periurban community practices, perceptions, and attitudes with respect to: the role of water in disease transmission, hygiene behaviour, diarrhoea its causes, transmission and prevention.

◊ To identify sources of water contamination.

The Study Area

Langas is a periurban area of the Eldoret Municipality in the Rift Valley Province of Kenya. Established as a municipality in 1958, Eldoret is currently the 5th largest town in Kenya with a population of about 150,000. Eldoret is located 65 km North of the Equator at an altitude of between 1800 m and 2750 m above sea level. The periurban areas around Eldoret became part of the municipality as urban growth stimulated the arrival of low-income workers and a migrant population. Eldoret is situated at an altitude which provides a temperate climate throughout the year. Langas is situated on a plateau 5 km south west of Eldoret town. In 1989 Langas had a population of 28,000 (Kenya census 1989-90), and the current population is estimated by intermediate rates as 37,500. The population density in 1989 was 220 persons per hectare. The main health care centres include Pioneer Health Centre and Eldoret District Hospital. The average family of seven occupies 1-3 rooms each measuring about 10 m^2. The houses are mainly constructed with mud and wattle walls (which may be plastered) and corrugated iron roofing sheets. The monthly room rents range between Kshs.150 and Kshs.500. [At the time of writing 1US $ = Kshs 55 Eds.]

The main sources of water are hand dug wells. A basic pit latrine is the commonest sanitary facility. Garbage is disposed of into open rubbish pits.

Methodology

Design. An 18-week longitudinal community based study of 39 mothers and their 124 healthy children aged between one and 12 years resident in Langas was conducted by 4 medical students.

Ethics and Etiquette. The study proposal was approved by the faculties Ethics and Research Committee. Recruitment was done by word of mouth in March and April 1994 with permission from the owners/caretakers of the selected plots. The nature, possible benefits, and inconveniences were fully explained and then informed consent was sought.

Sampling Technique. A two stage random sampling method was used to define a representative population of 39 mothers and 124 children from an estimated 6,000 children aged between one and 12 years. Random sampling of 8 plots was carried out at a predetermined distance from the bus stop. All plots had to consist of more than 10 contiguous households. When the selected plot did not meet this requirement, or where permission was denied, then the next plot which met the criteria was chosen.

From the total number of families in each of these plots ("Clusters of households") any consenting families with at least one eligible child were again randomly selected for interviews and detailed focused observations of behaviours related to water handling and sanitation. Out of a total of 46 eligible listed families, 6 families declined recruitment, leaving 40 families at the beginning of the study in May, 1994. For any family to be eligible it had to have no plans of migration during study period, and to have at least one child aged between 1 and 12 years.

The eligibility criteria for children were:

- they had to be between 1 and 12 years,
- they had to be fully immunized,
- they should have no history of chronic illnesses, or general ill health; and
- they should not be on any medication thought to affect general health or produce diarr-hoea as a side-effect.

Outline of experimental technique. On entry into the study the mothers were interviewed using a pretested semi-structured questionnaire to obtain identifying data, socio-economic and demographic characteristic; attitudes, knowledge, and practices as related to hygiene, water handling and diarrhoeal disease. They then gave a complete history of each eligible child. In addition during this first visit measurements of well-latrine distance, depth to well water level and actual depth of well and latrine were determined. Further the initial spot-check observations (using a structured observation protocol) and the extended-focused observations (2-3 hours) were carried out. Water samples were taken for baseline quality assessments.

Further interviews and observations were conducted fortnightly and each two week interval constituted one "round". On each subsequent visit the mothers were interviewed about the occurrence of diarrhoea in the preceding round. During the study, the students held 40 semi-structured questionnaire interviews and carried out 9 rounds of informal interviews and extended observations.

For the study purposes, diarrhoea was defined as the occurrence of at least 3 unformed stools in any 24-hour period during the round. Because of the difficulty in ascertaining when one episode ended and the next one began, it was arbitrarily stipulated that a child could have a maximum of one episode of diarrhoea in any one recall period.

Water samples were collected and analyzed for faecal coliforms by membrane filtration technique at the SWE laboratory at monthly intervals.

Two focused group discussions were conducted among women resident in the plots.

The observational findings led to the design of mini-studies (reported in poster sessions at this conference). Bacteriophage tracing was carried out to provide more conclusive data on the patterns of transmission and the effects of hygiene behaviour on water quality. Detailed description of these methods are found in the students COBES Report, 1994 (Library, Faculty of Health Sciences, Moi university).

Data Analysis

Initially 40 mothers and their 124 children had been enrolled into the study, however one mother and 7 children were excluded from the analysis because of violation of protocol which had been defined as missing more than one round of the study.

The mean, mode and median of the ages of the participants were determined using the EPI-INFO 5 programme. The incidence of diarrhoea was calculated by dividing the number of episodes of diarrhoea by the number of weeks of observations. The number of episodes of diarrhoea per child was also stratified by age of the children and the level of education of their mothers.

Analysis was carried out in three stages:

Construction of the Hygiene Index Criteria. The aim was to establish a numerical index for the hygiene practices observed in the compounds. The criteria devised a scale of one to five and took into account domestic waste disposal, latrine use practices, and faecal waste disposal. Thus a total plot score of three would represent the worst situation and 15 points would represent the best possible situation.

The hygiene index was then cross-tabulated with other variables such as bacteriological quality of water, mother's educational status and the incidence of diarrhoea.

Analysis of qualitative data. Categories of behaviours were defined according to similarity of answers and students judgement and led to three broad classifications of behaviours:

(a) Well-related

(b) Latrine-related, and

(c) Handwashing-related

A summary for each question or issue of discussion was outlined for the interview and focus group discussion data.

Laboratory data analysis. This was done for the results of water samples taken in baseline months of March and April, May August and September. Results for the June-July period were unavailable as the laboratory ran out of supplies. Colony counts were converted to faecal coliform counts per 100 ml and the average count per sample was determined. The average faecal coliform count for a well over the whole study period was calculated and designated as the bacteriological quality of water for that well. Counts recorded as confluent growth (CG) or too numerous to count (TNTC) were regarded as missing data due to the unreliabilty of the counts.

Variables. Variables were compared as follows; the bacteriological quality of water (BQW) to the latrine-well distance; the BQW to the incidence of diarrhoea in children; the BQW to the hygiene index, the mother's educational status to the incidence of diarrhoea; the hygiene index to the incidence of diarrhoea; and the incidence of diarrhoea to the age of the children.

Statistical Tests. Data were compared using the Chi-square analysis and Fischer's exact test. The tests were one-tailed, and statistical significance had been defined as p value >0.05 (5%).

Limitation of the Study

The study had disadvantages associated with a modest sample size, however considering the fact that the households were randomly selected, it is believed that the data are representative. As the study was limited to only a four-month period it is possible that practices affecting pathogen transmission occurring during other times of the year could have been missed. Finally it is presumed that the practices which were observed corresponded to the usual consistent behaviour but it is possible that some missed variations in practices could have contributed to individual episodes of diarrhoea. It is thought that the repeated nature of the observations were adequate to determine the consistent behaviours that would be representative of the hygiene behaviour in each plot.

It is important to acknowledge that the qualitative data were mainly observational associations that were likely to be influenced by personality, attitude, and diligence of the observer and other subjective factors. However, as students attended a two-week seminar on observational skills in the study of hygiene behaviour it is reasonable to say they were competent in the necessary skills to implement this kind of study.

There would, therefore be need for a controlled experiment to validate these observed practices as risk factors for diarrhoea.

Results

The age range of the 117 children was between one and twelve years with a mean of 5.75 years. The median age was 5 years. There were 59 girls and 58 boys. The mothers´ ages ranged between 16 and 47 years with a mean of 29.5 years and a median of 27.8 years. 50% of the mothers had completed primary school education and 37.5% (N=39) had completed secondary school education.

The general trend was that of an inverse relation between educational status and the incidence of diarrhoea. The highest number of episodes of diarrhoea was reported in the months of June and July.

Table 1. Mothers´ educational status compared to the incidence of diarrhoea.

Level of education	No. of mothers	Diarrhoea cases/child
Nil	03	3.00
Primary	19	2.30
Secondary	15	1.93
Tertiary	02	2.00

The incidence of diarrhoea is inversely related to the value of the hygiene index (3 worst and 15 best). This may imply that the better the hygiene practices within the plot the lower the incidence of diarrhoea. When the rates of diarrhoea were adjusted to the mean age of children per plot there was no significant differences (p value=0.05) after the adjustments.

The mean faecal coliform count was 300 faecal coliforms/100ml. There was no clear correlation between the well-latrine distance and the faecal coliform counts. The comparison of BQW and the hygiene index gave a curvi-linear correlation indicating a weak influence of hygiene practices on the quality of well water. This seems to imply that there are other confounding factors in this correlation.

Table 2. Relationship between the hygiene index and the incidence of diarrhoea per plot

Plot. No.	1	2	3	4	5	6	7	8
Hygiene index	6	7	5	6	7	8	11	6
Incidence of Diarrhoea	2.00	2.13	2.43	2.17	2.00	2.36	1.79	1.92

Table 3. Bacteriological quality (no. faecal coliforms/100ml) of well water by month and plot

Plot No.	Positive	May	August	September	Mean
1	00	00	63	1195	315
2	00	00	18	59	163
3	13	00	833	08	214
4	533	600	253	16	351
5	560	525	17	10	278
6	01	32	100	160	73
7	00	82	445	32	140
8	16	225	1900	1340	870
Average per month	212	183	454	352	300

Findings of the Focus Group Discussions

Two focus group discussions were held. They were convened by a facilitator whose role was to stimulate the participants until the subject was exhausted.

The general findings on diarrhoea indicate that there is a profound knowledge base on diarrhoea in the community. On the symptoms, frequent bowel movements "unusual to the person", watery and foul-smelling stools were among those mentioned. In one discussion two members talked about dehydration saying that the person "withers" and "the skin becomes dry and he looses weight".

The management of diarrhoea was common knowledge and included the use of rehydration fluids, both home made and from health centres. Certain herbal preparations were said to be effective in the management of diarrhoea. One discussion group brought up the use of ashes and charcoal in the management of diarrhoea.

The causation of diarrhoea was said to be due to bad food, unboiled water, teething in children, germs, malaria and being cursed.

On the mode of transmission of diarrhoea, one focus group noted that diarrhoea may be transmitted if a sick person contaminated food. Close contact with the person or his stools; utensils and flies were given as major transmission routes.

Results from Observations

Focused observations were made on three key hygiene behaviours that were thought to be important risk factors of diarrhoea and which are amenable to interventions: (a) safe disposal of faeces, (b) handwashing, and (c) keeping drinking water free from contamination.

The safe disposal of faeces is vital in reducing the spread of diarrhoea. At all of the observation points a pit latrine was the main sanitary facility. Most adults used the latrines, however, children under five hardly used the latrines at all. Children were frequently seen to defecate in the open playing grounds and their stools could be found all over the yard in five of the eight plots. A few mothers buried their children's stools. Most mothers believed that a child's stools are "harmless". Most latrines are poorly maintained and the poor state of latrines was associated with an increased presence of faeces in the yard and the surrounding open grounds. The cultural background of the people had an influence in a few cases where, for instance, they reported that children could not use the same latrine as their grandparents.

Handwashing is important since it is believed that hands are important vehicles of spreading diseases. Most of the residents of the plots were aware that hands should be washed after defecation and before eating. During these informal discussions people rarely mentioned the necessity to wash hands before preparing meals or after cleaning children's bottoms. It was noticed that mothers washed their hands whenever they contacted something they perceived as "dirty". But it was observed that quite a number of mothers did not wash their hands after cleaning a child's bottom.

The practice of handwashing was far less common amongst children and men. Handwashing with soap was very rare indeed. Unique episodes of handwashing included a lady using mud.

Mothers are hardly bothered by their children's hygiene practices and the consistent practice of handwashing was only observed in less than half the population.

Interestingly, it was common to see mothers using their kangas [decorative cloth often worn as an apron by women] to wipe their hands instead of washing them.

Protection of drinking water from contamination was the other key behaviour observed. The main source of drinking water in Langas is the hand dug well, the water kiosks in the area being either too far away, or too expensive for most interviewed residents. Faeces (human and animal) were sometimes seen uncomfortably close to the wells. The water drawing containers were frequently seen on the ground. Some land-lords and their tenants made attempts to keep their water sources safe. During the dry season wells were de-silted. Chlorination and the use of alum to treat the well water was observed but chlorination is erratic because the chemicals used are costly.

Observations at the home revealed the use of three kinds of containers for storage of water. The most common was a 20 litre plastic bucket and most of them were seen to be adequately covered.

Discussion

The general aim of this study was to determine in a longitudinal study whether the mother's perception, attitudes and practice of hygiene behaviour influenced the faecal coliform counts of well water and the frequency of episodes of diarrhoea in children.

The data indicate that hygiene behaviour is an important factor influencing the incidence of childhood diarrhoea in Langas. It was found that the incidence of diarrhoea was higher in those plots with a lower hygiene index performance.

It was found that the frequency of episodes of diarrhoea was inversely related to the children's age. Of the 191 episodes of diarrhoea reported, 73% (140) were in children under 5 who constituted 44 % of the population. The mean incidence of diarrhoea per child per 18 weeks was 1.6. Snyder and Merson (1987) in a similar study found an incidence of 2.2 episodes per child per year.

No correlation was found between well water quality and the distance to latrines. Our suggestion is that there are many possible confounding factors such as hydrogeology, hygiene behaviour and the presence of other sources of contamination.

The mean faecal coliform count at the wells was 300/100ml. It is to be recalled that Feachem (1984) had recommended that a source with faecal coliform count of between

10 and 100 could be used without treating the water. Van der Slice and Briscoe (1993) noted that water source contamination posed a significant risk for diarrhoea. They showed that if families using moderately contaminated sources (100/100ml) were able to use better quality water, diarrhoea among their children would be reduced by over 30%.

In the present study, correlation was found between the level of water source contamination and the frequency of diarrhoea in children using the water. A plausible explanation for this is that some of these children had developed immunity to the endogenous pathogens in the water, and that the possibility of asymptomatic enteric infections also exists. Earlier investigations by Barlett et al. (1985) and Black et al. (1987) support this possibility. In two studies of infants and toddlers in day care centres, new enrollees experienced significantly more diarrhoea than their playmates. After a few months of attendance the incidence of diarrhoea among the new enrollees dropped to the same level as the established ones.

Observations indicated that most latrines were insanitary, and that there is a possibility that the insanitary latrines could have resulted from poor hygiene behaviour by some residents; thus a vicious cycle started and propagated.

Handwashing before eating was uncommon and handwashing after defecation was erratic especially among children. Only 50% of the people washed their hands after using the latrine.

Most wells were not properly maintained but some visible attempts to protect water quality were observed.

From the focus group discussions it was clear that there is a vast and in depth awareness of the symptoms, transmission and causation of diarrhoea to such extent that ignorance cannot possibly be an explanation for the poor hygiene behaviours and the high morbidity due to diarrhoea.

The knowledge of diarrhoea was largely influenced by the socio-cultural background of the people and this must be considered in the approaches adopted by health care providers. Socioeconomic factors heavily influence hygiene behaviour and thus the general health of the community. As Briscoe put it in 1990 "people are not passive acceptors of threats to health but their actions to preserve health will always be secondary to those preserving life itself."

Conclusion

In this longitudinal study of 39 mothers and their 117 children, it was found that the incidence of diarrhoea was higher in the less hygienic plots, and that a hygiene index and bacteriological quality of water were substantially correlated. Thus hygiene behaviour influences the incidence of diarrhoea via multiple transmission routes.

Well water was heavily polluted by faecal matter and it was was almost universal, however, the found that knowledge of good hygienic behaviour actual practices were sadly minimal. This disparity between awareness and observed practices was attributed to the influences of socio-cultural values and economic determinants.

We suggest, then, that while current strategies aimed at reducing the morbidity and mortality associated with diarrhoea should be augmented, new interventions must be shaped by serious attention to socio-cultural values and economic determinants of health within the community.

It is important to create a socially prudent health care provision approach, including efforts to develop preventive strategies that put greater emphasis on moulding the existing socio-cultural norms to yield a better health for all.

Acknowledgement

This paper is based the field work of medical students in the COBES project, under the author´s supervision. The students involved were Asheesh Aggarwal, Patrick Sawa, James Kamau and Lulu Oguda. Their tremendous effort is highly appreciated.

References

Aung, Myo Han & Hlaing, T. (1989) Prevention of Diarrhoea and Dysentery by Handwashing. *Trans. Royal Society of Tropical Medicine & Hygiene*. Vol.83;pp 128-131.

Barlett, A. Moore, M. et al. (1985): Diarrhoea illness among infants and toddlers in day care centres, Z. Kinderheikd Vol.107, pp 495-502.

Black, R.E. et al. (1981): Handwashing to prevent diarrhoea in day care centres, *Am, J. Epid.*, Vol.113 pp445-451.

Boot, M. and S. Caincross. (Eds). 1993*: Action Speaks: The study of Hygiene Behaviour in water and sanitation projects*. The Hague: IRC.

Briscoe, J. (1984): Intervention studies and the definition of the dominant routes, *Am. J. Epid.* Vol. 120, pp 449-455.

Briscoe, J., J. Akin & D. Guilkey. 1990. People are not passive acceptors of threats to health endogeneity and its consequences, *Int. J. Epid.* vol. 19, pp 147-153.

Caincross, S. 1992. Control of enteric pathogens in developing countries. In *Environmental Microbiology*, Mitchell, R (Ed). New York: John Wiley, pp 157-190.

Caincross, S. and R. Feachem. 1983. *Environmental health engineering in the tropics: An introductory text.* Chichester: John Wiley.

Clemens, J.D. and B.F. Stantorn. 1987. An educational intervention for altering water-sanitation behaviours to reduce childhood diarrhoea in urban Bangladesh. *Am. J. epid.* Vol.125, pp 284-291.

Feachem, R.G., Bradley, D.J., Garelick, H. and D.D. Mara. 1983. *Sanitation and diseases: Health aspects of excreta and waste water management.* John Wiley, N.Y.

Han A.M., and T. Hlaing. 1989. Prevention of diarrhoea and dysentery by handwashing. *T.R. Soc. Trop. Med. Hyg.* Vol.38, pp 128-131

Huttley, S.R.A. 1990. The impact of inadequate sanitary conditions on health in developing countries. *Rapp. Trimest. stat. Mondiale.* Vol.43.

Huttley, S.R.A. et al. 1990. The Imo state drinking water supply and sanitation project 2. *T. R. Soc. Med. Hyg.* vol.84 pp 316-321.

Khan, M.U. 1982. Interruption of shigellosis by handwashing *T.R. Soc. Med. Hyg.* Vol.76, pp 164-168.

Shuval, H.I. et al. 1981. Effect of investments in water supply and sanitation on health status. *Bull. WHO* Vol.59, pp 243-248.

Snyder, J.D. and M.H. Merson. 1982. The magnitude of the global problem of acute diarrhoeal disease. *Bull. WHO* Vol.60, pp 612.

VanderSlice, J. and J. Briscoe. 1993. All coliforms are not equal. *Water Resources Research,* Vol.29, No.7.

White, G.F., Bradley, D. and A. White. 1972. *Drawers of water*. Chicago University Press, pp 162-176.

COMMENTARY ON *Theme II*

Mr. S. Mgana (Ardhi Institute, Tanzania) pointed out that water quality varies greatly from one place to another. He asked if there was any relationship between latrines at different levels in the aquifer and wells in the area.

Mrs. Watts (Moi University) mentioned studies in Lusaka and Harare in which pit latrines were found to spread contaminants despite the use of piped water. She posed the general question, how can sanitation be made sustainable?

Ms. Elena Hurtado (Institute of Nutrition of Central America and Panama) asked why medical and social sciences are kept separate in these studies. They could have been better coordinated. Undertaking such studies creates expectations that are difficult to fulfill. Do any of the presenters have solutions to offer?

Mr. Stenström (Swedish Institute for Infectious Disease Control) rose to clarify the intent of the study. It was not aimed at intervention but just to promote understanding. The wells are uncovered and subject to pollution from both ground and surface sources. Tracer studies of groundwater showed 130 metre transport of low concentrations of faecal coliform bacteria from latrines to wells within three days. Safe distances between the two are not respected.

Mr. Akong'a replied to the questions and comments by noting that he was not sure how sustainability can be managed. The dynamics of municipal councils, the varying rules and guidelines for dug wells and for emptying waste all figure into the arrangement. Medicine and social science were kept apart in the presentations but there were preliminary discussions between anthropology and medical participants before the questionnaire was finalized, and every week during the study there was a presentation by one or another participant. At all times health issues were kept foremost.

Mr. Kamar referred further questions about pollutants to *Mr. Stenström's* presentation of the afternoon (paper 9).

THEME III : WATER QUALITY AND INDIVIDUAL CHOICE OF WATER

Chairman: Joshua Akongá

Paper 6. LOCAL AND PROFESSIONAL PERCEPTIONS OF WATER QUALITY IN KENYAN PERIURBAN SETTLEMENTS: LANGAS AND HURUMA

Richard M. Swiderski
956 Old Quarry Road South, Larkspur, CA 94939 USA

A study of water supply and water use in Langas and Huruma, two settlements near Eldoret, Kenya, offers a basis for speculation on the reasons water-borne pathogens affect the community as they do. A quantifiable relationship between water supply and morbidity is difficult to gauge even with protracted study because of the fluid nature of the population, because households vary greatly in composition and because children may not center their activities around any one water source. I chose a method that combined observation with casual conversation and interviews in an attempt to obtain guidance for further more detailed investigation. My aim was to adduce community perceptions of water quality as a groundwork for further study.

I began with the proposition that natural selection in favour of relative immunity and acquired immunity through exposure operate in tandem with cultural adaptations seen in water choice and water use, and that biology and culture are interrelated in ways that an evolu-tionary hypothesis may help us detect. I posited that the presence of particular pathogens, the opening and closing of peculiar transmission routes and other influences (climatic, dietary and political) on the health of people all play significant roles in the health profile of the population, and that they are directly and indirectly related to individual response to infection.

Neither water use perceptions nor individual biology has been carefully assessed in these populations, so I decided to initiate the question by defining the behavioral dimension. Close long-term observation is required to take account of water consumption behaviour. This rough sketch only includes general points that may substantiate or refute an evolutionary thesis: that water use patterns are generally are adaptive to excluding pathogens but may fail and even help transmit and amplify virulence when underlying conditions, population, character of the water supply for instance, undergo a change.

Municipal water supply to Langas and Huruma is through a system of public taps connected to the urban system. This water is safeguarded by treatment routines and regular testing. But households do not obtain all of their domestic water from the public taps. Other sources, the river, wells and rainwater collection points, may be more convenient to individual groups of households. Commercial water vendors obtain water from taps and carry it to the more remote dwellings. There is a choice of water sources most of them not subject to preventive treatment. Bottled water is not available, but bottled soft drinks, beer and locally fermented brews account for a small portion of liquid consumption.

A significant amount of the daily water requirement is taken in the form of cooked food and tea. However water is obtained the early morning routine in many household clusters is to boil water and use it to make uji, a millet porridge, or ugali, a maize meal stew accompanied by greens and rarely, meat. Ugali, once prepared, can be kept for long periods of time and consumed in different forms. There is little direct drinking of water. At the very least it will be taken in the form of tea. The other major contact with water is washing face, hands and body and rinsing out the mouth. This is raw water taken from the storage vessel. The taps and wells are for getting water but not for using it then and there.

Water for washing is kept separate from water for cooking and drinking in most households. They come from the same sources, but the washing water is not boiled or treated before being used while the cooking water always enters the body through a medium, food or tea, that has been heated. There is a strong distinction between new water and used water. New water is stored in its own containers; used water is allowed to flow away through drains, over the ground or is used to water plants. The main water assessment criterion is the distinction between new and used water. New water is from a fresh source and meets variable smell and flavour criteria; used water is dark in colour, has a distinctly foul ("dirty") odour and is greasy or soapy. Used water emerging from a new source is rejected. Used water cannot be remade into new water by heating or by any other treatment. Any new water which falls on the ground or into a used water container becomes used.

Tap water and well water are not new by source alone, but must meet the same criteria as water from any other source. There are reasons to believe that water from these sources might be used. For instance, it shows fluctuations in odour, taste, and appearance often explained by rains, winds, and other weather phenomena. Tap and well water then changes as other new water, river water, also changes. Changes unrelated to weather, for instance the sharp change in flavour when fluoridation or chlorination compounds are added to the tap or well supply, arouse considerable suspicion. The water then seems to be used. The people with first access to the water are passing along the water they used to carry away their chemicals. "Water etiquette" extends to commonly used well water simply because it is to be used in common but it does not extend to river water. Only an enemy from another community poisons a well while anything can be thrown into the river without a thought about those downstream. Well water's newness is a shared responsibility while the river is always new. Well water is more likely than tap water to remain new with radical fluctuations of quality because the causes of these fluctuations are tangible. When the rains come and the water rises in the well or the opening is flooded, this is only to be expected. The water is no less usable than before. Water is kept new by the covering of earth. The flow of water beneath the earth is a heartening concept to community members. It means that well water is like river water after all.

River water is least likely to become used because it flows constantly and is forever renewing itself. It hardly matters that the flow carries factory and human waste, silt and detritus from upstream. The flow in the immediate vicinity of water gathering is all that is needed to renew it. Christian groups hold total immersion baptisms in river pools ignoring the dead cow and the discarded oil filters a little ways upstream. Flowing water purifies what it touches.

Only during the dry season, when the "stony river" that gave Eldoret its name shows itself, do people hesitate to dip into the river. During that time of scarcity the well and tap water are even more likely to seem used because ”people need water now” and what water there is must already be used.

Vessels are important in water assessment. Preferences range from pottery to plastic and metal. Discarded glass bottles also are used. It is assumed that in each household certain vessels hold only new water, and that water in these vessels can be used for personal washing and food preparation. In most households there is no vessel specifically for drinking water. Water is boiled in the same container for tea and for food, though some households keep a separate tea kettle or tea water pan. The three-footed clay pot which has long been the centre of consumable preparation is still the centre of some households. These vessels are difficult to obtain and are replaced by metal pots when they break.

The ability of the pottery vessel to keep water cool during hot weather is widely recognized and this by extension gives water from pottery some medicinal efficacy against fever.

Across a number of households and through time the stages of water use behaviour become visible. Drinking new (no less used) water or eating raw food are viewed as lapses to a more careless way of conducting life. It is assumed that there was a time when people did not process water by cooking, and that was a time of serious disease and general misfortune. That time in the past when people learned to process water but still could farm the land as families and hunt freely was happier than the present. Bottled drinks and water from indoor taps are advanced and superior forms of water, better tasting and less likely to contribute to ill health. They are the newest, and most expensive, water.

The rainy season, when even the tap water can run muddy due to great flows of loose topsoil carried into the streams, is the primitive time of water consumption. Then extreme care has to be exercised to maintain the separation between new and used water. Soil is allowed to settle out before water is used: only animals and barbaric people drink muddy water. Herbs can be placed in muddy new water to settle it and sweeten the taste. Some people distinguish flavors new water imparts to food and treating it herbally is a preliminary to its use in cooking.

New water maintains health though it is not clear how. It can also restore health by turning into used water and carrying off whatever substances have been troubling the ailing person. This is most dramatically visible in the treatment of diarrhoea, which is literally washed away with new water. The water carries the diarrhoea infection but the clothes and the hands of the washer do not. Diarrhoea goes away with water but it does not come from water. It comes from food that has not been prepared. Diarrhoea is a flow of bad matter out of the body which can be encouraged with new water washing. Diarrhoea flow is the most used water.

Women are the main gatherers and preparers of water, and the keenest in articulating the new-used distinction. Men may drink unprocessed new water or even used water and feel that it proves their strength if there are no consequences. Children have the habit of eating whatever they can put their hands on, and they slowly learn to make distinctions. Children suffer diarrhoea and stomach upsets more often than adult men or women because children have not learned to prepare what they take into their bodies and because they do not know how to encourage the flow with new water. Their mothers or older sisters must wash them.

Since the new water that enters a household is routinely consumed through cooked food and beverages all members of the household take in the same water. Children are in the process of adapting to the particular mix of organisms that succeed in entering them through normal consumption. Because they require a greater flow of water children are more often exposed to untreated new water and may themselves consume used water because they do not distinguish new from used water. Children's more frequent intestinal and skin ailments are part of their hardening into an adulthood then relatively free of these problems. Children learn the new-used distinction from practice, from associating health or illness with certain kinds of food and water. A child's failure to recover from a typical sickness usually brings recourse to directed healing techniques to make up for what the child lacks or has lost. Healing often employs specialized forms of new water.

Water supply rather than water quality is the chief articulated concern in Langas and Huruma. If supply is sufficient then quality will be adequate. If supply is low, if the river is rocky, the wells are deep and the tap water does not run constantly, then the newness of the water is in question. Some households are more particular than others about water quality, which means they will seek water from other sources (wells, pools, taps outside the neighbourhood) when they are not satisfied with the immediate sources.

This opens lines of pathogen transmission which may be invisible to anyone from the out-side seeking the causes of an outbreak. The outbreak comes to the attention of outside specialists (if it ever does) when the normal adaptations inherent in the graded assimilation of typical pathogens from childhood onward, and the complex of new versus used water breaks down, or a new highly virulent pathogen enters the system and cannot be assimilated. Events leading to change in water use behaviour also can affect the spread of water borne pathogens. Faced with shortages of water they would deem acceptable periurban residents will use sources at hand against their own better judgment. In this they are like everyone else, except that their judgment is finely developed and they will choose the best water available. There are widespread local perceptions of water quality but no one source is optimum. In other words, perceptions of water quality are not scientifically defined and thus are not realized in a single controlled source. There are many waters and no one kind of water for the people of these settlements.

The irregular management of actual sourceing in the communities is incomprehensible to professional water analysts, public health officials and water managers, even those who are quite familiar with life in the settlements. They have accepted the concept of strategically controlled water sources and take it for granted that all residents will want to have access to health-securing supplies. The main disparity between professional and local perceptions is that residents of Langas and Huruma do not equate formal water purity (for instance, low coliform bacteria count) with health and do not believe that water that has passed through the pipe system is by definition better than any other water.

Water quality comes under professional attention only routinely, as part of monitoring the piped water supply. For the most part public health officers do not think of analyzing community water until there are clear signs of water-borne disease. As it was in Europe when water treatment was first introduced during the latter 19th century, scientific analysis of water is directed toward improving the quality of end-of-pipe water. In communities where piped water is the predominant source, the people's own adaptation prevents common infections detectable by tests from establishing themselves.

Periurban communities like Langas and Huruma present a challenge to this public health policy. They are at least as densely populated as the urban centres but their water intake cannot be controlled nor can the quality of water be checked from water passing through a treatment plant or from samples taken from taps. Professionals do not take account of the many water sources and the manner of use in these communities, and they are unaware of the new-used water distinction.

There is a received opinion that increased sophistication of laboratory analysis, better equipment and more scientifically trained staff will eliminate the threat of water-borne disease. There is also a vague belief that public health education, which means education in the scientific definition of water purity, will improve the health of periurban people. Yet these people "do not get sick more often" because their water seeking and water use behaviour are refined in ways different from those prescibed by hygiene and effective in ways that do not fit traditional scientific categories yet are effective just the same. If the residents of Langas and Huruma acted simply in violation of hygienic rules they would get sick more often. In fact, they follow other rules.

Professionals think of water as a chemical entity and a bacteriological medium. They sample water at specific points in the community but they do not ask community members where they actually find their water. And they do not take the time to observe where water actually is collected, how it is stored and treated, who consumes it, in what quantities and when.

The scientific perception of water quality unfortunately precludes the ethnographic. It is difficult enough to keep track of disease incidence in these communities. It has practically become a cliche of epidemiology that an epidemic is not noticed unless it starts or spreads in a hospital and health workers are affected. If water analysts were to take their task one step further and analyze blood samples from the community for antibodies to common water-borne diseases they would find a history of infection that the limited morbidity records have never disclosed.

An ethnographic examination of water quality perceptions in the community begins to show a complex pattern of behavioral adaptation to the commonest diseases. Individual cases of infection are either part of the adaptation process or are a sign that the adaptation has failed for this one person.

I did not have the opportunity (and do not regret it) to observe in Langas and Huruma reactions to cholera, typhoid and other diseases caused by water-borne pathogens. I am convinced, however, that specific household water use routines inhibit the spread of pathogens and keep virulence at a low level.

It also appears that these same routines allow if not encourage the level of infection which does prevail. This may maintain a level of challenge to the immune system of each resident. When the household population density rises above a certain level, especially as the result of an influx of refugees, who put a strain on all existing resources, there is every chance for water-borne pathogens to leap the behavioral gaps and take on a more virulent form. The pathogens adapt to take advantage of oppor-tunities human present. With a more detailed account of water use in periurban households it will be possible to see where these oppor-tunities arise.

It is cruel for anyone, particularly for an outsider who is not resident in the settlement, to say that even a single case of diarrhoea in the long run benefits the whole. This is the thinking that has long given evolutionary theory a bad name. I conclude with a call for more detailed research on household water acquisition and use and for careful thought on the ethical implications of any policy that is formulated as a result. Understanding the difference between professional and local perceptions of water quality is the starting point for any research no matter how scientific. The researcher must be conscious of how his or her scientific biases may cause the most significant information to be obscured or dismissed. Above all it is neces-sary to promote an analytic perspective among the people themselves.

Paper 7. WATER QUALITY, QUANTITY AND HEALTH

Sandy Cairncross, London School of Hygiene and Tropical Medicine, Keppel St., London WC1E 7HT, UK

The Bradley classification

Much of the discussion at this Conference has focused so far on problems of water contamination, and water quality, with the implicit assumption that improved water quality will bring better health. I want to question this assumption and set it in a broader context of the health impact of water supplies, but first I should set out the basics of the Bradley classification of water-related diseases (White, Bradley & White 1972; see Table 1).

The most important aspect of Bradley's classification for water supply planners was his distinction between **water-borne** transmission, in which a person is infected by drinking water containing the pathogenic organism, and **water-washed**, perhaps better termed **water-scarce**, transmission which occurs when not enough water is available for personal and domestic hygiene. Other diseases are **water-based**, being transmitted by aquatic hosts, or transmitted by **water-related insect vectors.**

Table 1. The Bradley classification of water-related disease transmission routes

Water-borne transmission	You drink it
Water-washed (or water-scarce) transmission	Person-to-person transmission due to lack of water for personal and domestic cleanliness
Water-based transmission	Aquatic intermediate host (e.g. a snail)
Water-related insect vector	Transmission by insects which breed in water or bite near water

However, this is a classification of transmission routes, not of diseases. For that, a slight reorganisation is required, because one group of diseases, the faeco-oral infections, can be transmitted by both the water-borne and water-washed routes (Table 2).

Table 2. Adapting the classification to diseases

Water-borne	Faeco-oral
Water-washed	Skin and eye infections

This group, which includes the diarrhoeal diseases as well as typhoid fever and hepatitis A, is the group where the impact of improvements in water supply is likely to be of the greatest public health importance; other diseases kill fewer people or are less likely to be affected by water supplies. An additional group, the skin and eye diseases, is exclusively water-washed. The potential of the faeco-oral infections to be both water-borne and water-washed raises an important question for the water engineer: which type of transmission route is more important? Should one spend one's scarce resources providing water of the best quality, and so prevent water-borne transmission, or should they rather be spent on making water more accessible to people and increasing their use of water for hygiene, thus reducing the amount of water-washed transmission?

Problems of measurement

Before I try to answer this question, I must explain the difficulty of measuring the effect on health of improvements in water supply, and hence the difficulty in interpreting the literature to reach any firm conclusions regarding the specific improvements which are beneficial to health.

Rather than entering into a discussion of epidemiological theory, let me illustrate the problem with an example; a remarkably thorough study conducted in Tanzania by a Danish team as part of a study to produce a master plan (Prag *et al.* 1983). They compared two villages, Namabengo and Mkongo, near Lake Niassa. Most people in Namabengo drank water of good quality from a piped water supply, while the main source of water for Mkongo village was a heavily contaminated dug hole.

Table 3 shows the proportion of children under 5 years of age in each village whose mothers reported that they had suffered from diarrhoea in the previous week, and the proportion with a serological indication that they had been infected with the typhoid fever bacillus. It is clear that children drinking the dirty water from the dug hole in Mkongo were more than three times as likely to have diarrhoea or to be infected with typhoid than those drinking the clear piped water of Namabengo.

Table 3. Results of a health impact study for children < 5 years

	Diarrhoea in last week		**Positive widal (S. typhi H. antigen)**	
Village	Piped water	Dug hole	Piped water	Dug hole
Namabengo	15/216 (7%)		4/216 (1.9%)	
Mkongo		39/134 (29%)		12/133 (9.0%)

Source: Prag *et al.* (1983)

And that would be all; one would conclude that Namabengo's water supply had an excellent impact on child health, were it not for the fact that some families in Namabengo drank water from a dug hole at one end of the village, while a minority of residents in Mkongo had access to a piped water supply. Adding these people to the table changes the picture considerably (Table 4). It becomes clear that it is the village where one lives, not the water one drinks, which affects the risk of disease transmission.

Table 4. The **complete** results of the health impact study for children < 5 years

	Diarrhoea in last week		**Positive widal (S. typhi H. antigen)**	
Village	Piped water	Dug hole	Piped water	Dug hole
Namabengo	15/216 (7%)	5/70 (7%)	4/216 (1.9%)	1/70 (1.4%)
Mkongo	37/100 (37%)	39/134 (29%)	7/100 (7.0%)	12/133 (9.0%)
Total	52/316 (16%)	44/204 (22%)	11/316 (3.5%)	13/203 (6.4%)

Source: Prag *et al.* (1983)

To understand why this should be, one needs to know a good deal more about the villages. For example, there was a large mission hospital in Namabengo where the staff handed out free antimalarial tablets for all the children; the Danish researchers also found that children with malaria parasitaemia were also more likely to have diarrhoea, which is plausible as the malaria parasite suppresses the host's immune system as a self-protection measure. One would also need to know that there had recently been an epidemic of typhoid fever along the shore of the lake, which had reached Mkongo village but never gone up to the higher, cooler regions where Namabengo was situated.

This should be enough to make you sceptical of many epidemiological studies. In fact there have been about a hundred such studies of the health impact of water and sanitation in recent decades, and practically every one of them has important methodological weaknesses (Blum & Feachem 1983). Interpretation of the literature is thus a matter of personal judgement.

An illustrative epidemic

The interpretation I am now going to offer you is therefore based on my inevitably subjective assessment of the literature. But I am happy to say that it is also the view of the World Health Organisation's specialists in diarrhoeal diseases control. It is this:

> **We all know that faeco-oral disease can occur in water-borne epidemics, but most of the endemic transmission which occurs in poor communities where hygiene is almost inevitably poor is in fact water-washed.**

Rather than present a conclusive study to demonstrate this, I think it is preferable to illustrate it by an anecdote. in 1976 Kosti was a small town on the Nile in Sudan, with a population of 65,400. The water supply abstracted water from the Nile, at a point just below the jetty where the Nile steamboats used to dock. Excreta from the crew and passengers on these boats were discharged to the river, and one can presume that a fair proportion of the abstracted water consisted of dilute sewage.

Now, this need not have been a problem in itself, as the water was treated and disinfected with chlorine. After all, most of London uses water from the Thames, which at the end of a dry summer (such as this year) contains over 30% sewage from the upstream towns such as Oxford, Reading and Marlow.

But one day, they ran out of chlorine. The result was a typhoid epidemic in the town. Figure 1 shows the epidemic curve (Elzubier 1977). The chlorine ran out in early July. Typhoid has an incubation period of 7 to 14 days. Sure enough, the epidemic began within a week or two. Within days, a team of Sudan's best epidemiologists rushed to Kosti to investigate. By the 21st they had traced the epidemic to the water supply and instituted preventive measures, particularly the reinstatement of chlorine disinfection and the regular examination of the water for chlorine residuals and faecal coliforms.

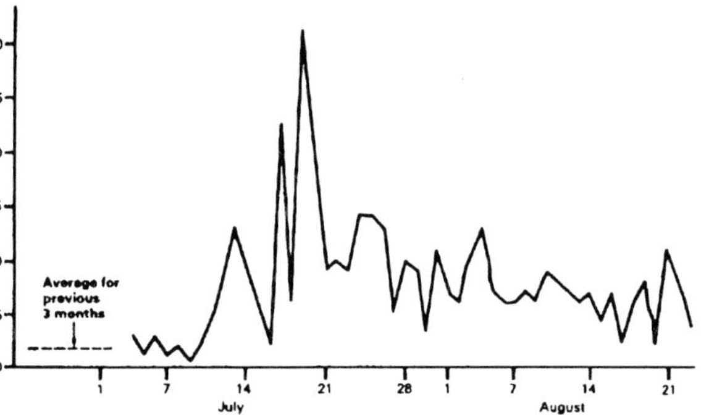

Fig. 1 Typhoid epidemic curve in Kosti

Notice, however, that the cases of typhoid continued to come in, for much more than 14 days after the remedial measures had been put in place. We can be certain that these cases were not waterborne; in the circumstances the disinfection of that water supply and the quality of the water were being monitored extremely carefully to ensure that water-borne transmission would not occur. These cases must therefore have been water-washed.

In fact, typhoid had not been absent from Kosti before the epidemic, as can be seen from the graph. Of course, there had to be at least one case to start the epidemic off. The fact that there were more cases in the period after the epidemic than immediately before is understandable; after the epidemic there were more temporary carriers, people who had been infected and who continued to excrete *Salmonella typhi* in their faeces and contaminate their environment with it, creating the conditions for non-waterborne typhoid transmission. After the epidemic, of course, health workers were also more alert to typhoid cases and thus more likely to detect them.

Taking the epidemic proper to have lasted, say, 20 days with a mean of about 12 cases per day, we can say that it included up to about 250 cases of water-borne typhoid. However, if we assume that the previous three months were typical of the previous year, with an average of about 1.5 cases/day, then it follows that there would have been over 500 cases of endemic typhoid in Kosti during the year; twice as many as occurred in the water-borne epidemic.

It goes without saying that no public health official rushed to Kosti to investigate these endemic cases; or thought to write about them in a dissertation. The example helps to illustrate why we tend to give more emphasis to unusual, unexpected and threatening events such as epidemics, than to the constant but sporadic occurrence of individual cases of disease which are the day-to-day reality of the poor. Besides, the operator of a water supply can be held responsible for a water-borne outbreak; but only the victims are likely to blamed for the endemic cases whose immediate cause is inadequate hygiene.

The water use plateau

If the endemic faeco-oral diseases are mainly water-washed, under what circumstances will water supply improvements help to control them? In the discussion which follows, water consumption is taken as a proxy variable for hygiene. Of course, some households may use large quantities of water but have poor hygiene, while others may manage wonderfully with only tiny amounts. White *et al.* (1972) found that ethnic groups which customarily used large water containers tended to use more water than others. However, if we consider a single population, and a water supply has no effect on their domestic water consumption, it is unlikely to have had any effect on their hygiene. On the other hand, if it leads to an increase in their water consumption, the evidence suggests that much (probably most) of the increase will be used for hygiene purposes (Cairncross and Cliff, 1987).

When White *et al.* (1972) studied the relationship between water consumption and distance to the water source, they found a remarkable and unexpected result. Domestic water consumption was quite independent of distance over a range from a few hundred yards to about a mile. Since then, studies in various countries of Africa, Asia and Latin America have confirmed that households for which water collection involves a round trip of thirty minutes consume no more than those needing only three minutes. Within this range, they are on the water use plateau (Fig 2). As one might expect, households whose water sources are farther away use less, and those with access to water on their plot or in the house use very much more.

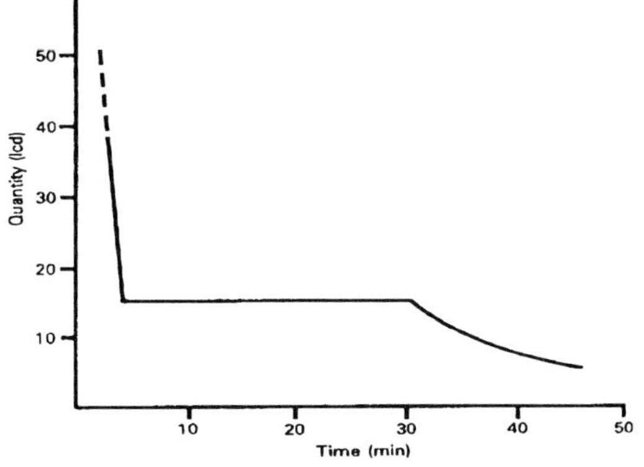

Fig. 2 The water use plateau

Here then we have a model to predict when a water supply improvement is likely to lead to an increase in water consumption, and hence an improvement in hygiene and a reduction in water-washed disease transmission. This will happen if one of two conditions is fulfilled:

(i) the old source is more than half an hour's round trip away and the new source is closer; or

(ii) the new source is on the plot, or at least within 3 minutes' round trip from the home.

This targeting policy will also ensure a reduction in other water-washed diseases, give the maximum benefit in terms of savings in time and drudgery spent collecting water, and increase the likelihood that consumers will be willing to pay for the water provided (Cairncross 1988).

Further evidence for non-waterborne transmission

Some people, having spent a lifetime believing, for instance, that typhoid is a waterborne disease, may find these ideas too radical. Let me therefore briefly summarise the evidence for the non-waterborne transmission of endemic faeco-oral disease. Briefly, it is of four main types.

Negative health impact results. A number of studies have failed to detect a health benefit from a water supply improvement. Many of these have involved water quality improvements alone, or improvement in quality plus improvements in access which only move households along the plateau, and therefore did not lead to increased water consumption (Cairncross 1990). A few studies, in which it has been possible to separate the effects of water quality and water availability, have found the health benefit associated with the latter but not the former (Victora *et al.* 1988). These results are entirely to be expected if the disease was mainly water-washed.

Food microbiology. Microbiologists are often shocked to discover hundreds of faecal coliforms in 100 ml of water from the storage vessel of a poor household. They would have been still more shocked to find thousands or even hundreds of thousands in every **gramme** of weaning food in the same household, and yet such levels are typical of those found by the few microbiologists who have looked - especially when the food has been left overnight, allowing the bacteria to multiply. Moreover, any faecal pathogens on food are likely to be swallowed together, producing an infectious dose; pathogens in water are more likely to be dispersed. They also do not multiply as they have no substrate.

Seasonality of diarrhoea. The following type of evidence is more speculative and circumstantial, but it deserves mentioning. In those settings where there is wide seasonal variation in temperature, the incidence of the bacterial diarrhoea and typhoid tends to peak in the warm season, whether it is wet or dry. The viral diarrhoea peak in the cold season.

In Western Europe and North America, the seasonality of diarrhoea changed during the first half of this century; from being a summer complaint, often of bacterial origin, it became a winter disease, largely caused by rotaviruses. The chief difference in the transmission of bacteria and viruses is that bacteria can grow in the environment, while viruses cannot. Bacteria need a warm temperature to multiply if they are on a nutritious substrate, while viruses need a cool temperature to survive as long as possible. If bacteria are not on a substrate, they also need a cool temperature to slow down their metabolism and prolong their survival. From this one can infer that, at some stage in their transmission cycle, these summer diarrhoea have had a chance to multiply; that is, they have been on a substrate such as food, and not in water.

Hand washing and fly control studies. The fourth type of evidence is the results of studies where interventions to promote hand washing with soap (e.g. Khan 1982), or the control of domestic flies, have brought about significant reductions in the incidence of diarrhoea, which would hardly have been achieved if most of the transmission was water-borne.

Conclusion

It would be quite wrong for the reader to draw the conclusion that the quality of drinking water is unimportant. Of course it is important that water supply agencies strive to provide water of adequate quality to ensure that water-borne epidemics are prevented. This is particularly important when large numbers of people are served by the same supply, which could cause a major outbreak if it were contaminated. However, we need to have more realistic expectations of the health benefits than some have had in the past, and to avoid extravagant and impractical measures such as the disinfection of every well in a shantytown, giving preference to an effective targeting policy to improve access to water for as many households as possible.

References

Cairncross, S. (1988). Domestic water supply in rural Africa. In: *Rural Transformation in Tropical Africa*, D. Rimmer, ed. London: Belhaven Press.

Cairncross, S. (1990). Health impacts in developing countries: new evidence and new prospects. *Journal of the Institution of Water and Environmental Management 4*, (6), 571-577.

Cairncross, S and Cliff, JL (1987). Water use and health in Mueda, Mozambique. *Transactions of the Royal Society of Tropical Medicine and Hygiene 81*, 51-54.

Elzubier AG (1977). Diploma in Tropical Public Health Dissertation. London School of Hygiene & Tropical Medicine.

Khan MU (1982) Interruption of shigellosis transmission by hand washing. *Transactions of the Royal Society of Tropical Medicine and Hygiene, 76* (2), 164-168.

Prag JB, Balslev K, Boesen J, Kapinga BS (1983) *Water Master Plan of Iringa, Ruvuma and Mbeya Regions, Tanzania* vol. 13 Ch. 11.

Victora CG *et al.* (1988) *International Journal of Epidemiology*, 17 (3) 651-654.

White G, Bradley DJ and White AU (1972) *Drawers of Water; Domestic Water Use in East Africa*. Chicago University Press.

Paper 8. DIARRHOEAL DISEASES AMONG THE CHILDREN, PEOPLE´S PERCEPTIONS AND HEAT TREATMENT OF WATER IN A COASTAL RURAL AREA OF KENYA

Mohamed Karama, Yoshio Iijima, Joseph Oundo and Peter Waiyaki,
Kenya Medical Research Institute, P.O. Box 54840, Nairobi, Kenya

Introduction

The epidemiological study presented here focusses on environmental hygiene and reduction of diarrhoea, with an emphasis on water quality. It was conducted in the coastal area of Malindi and comprised three phases; examination and etiological analysis of stool samples from all the cases with diarrhoea among under five admitted to the hospital of Malindi, collection of demographic base-line data for the villages from where the diarrhoeal cases came in order to identify the infectious routes, and identification of the traditional perceptions of diarrhoea. Finally, we started a trial to prevent diarrhoea by heat treatment of drinking water.

Village Water and Its Quality

Although Malindi town has piped water connnections, the periurban communities use river, pond, well and sometimes roof catchment of rainwater. These settlements have no modern sewerage systems. The villagers use pit latrines, and some of the pits reach the ground water. This may lead to contamination of well water with high coliform counts (see papers 13 and 14 in the preceedings). River and pond water were also contaminated with the same levels of coliform. Rain water, which is sometimes used by some villagers was not contaminated by coliform bacteria, but other bacteria. The majority of diarrhoeal patients were from these periurban communities.

Bacterial contamination of water. We visited five areas from where the Malindi hospital patients came and collected water samples stocked in their houses. The bacterial load in the water was tested. The water which were used for drinking were highly contaminated by bacteria. Bacteria concentrations of 10^4-10^5 CFU/ml and more than 10^5 CFU/ml were detected in 53 % and 21 % of the households, respectively. Low concentration of bacteria was found in a few samples.

Coliform bacteria at the concentration of more than 10^2 CFU/ml was detected in 72 % of the households, and more than 10^4 CFU/ml was found in 8% of the households. The bacteria isolated from the water belonged mainly to the following species: *Escherichia, Aeromonas, Serratia, Proteus, Streptococcus, Chromobacterium, and Vibrio. Salmonella spp.* was isolated from one sample only.

Identification of Causes of Diarrhoea

Causative pathogens. Diarrhoeal diseases are major causes of morbidity and mortality among children in the developing countries. Diarrhoeal diseases account for an estimated 12,600 deaths each day in children in Asia, Africa and Latin America. Diarrhoeal diseases, which are mainly caused by microbes including bacteria, parasites and viruses, can be caused by more than 20 microbial agents. While pathogenic *Escherichia coli* and rotavirus have been esti-mated to predominate in the developing countries, *Vibrio cholerae* and *Shigella spp.* sometimes causes serious outbreak.

Stools were collected from patients under five years of age who attended Malindi

Hospital (in- and outpatients) between July 1991 and June 1993. A total of 862 specimens were tested for bacterial and parasitic pathogens and approximately half (427) of them were tested for rotavirus.

The causative agents of diarrhoeal diseases were as follows: rotavirus (16 %), EPEC (9 %), *E. histolytica* (8 %), *Salmonella spp.* (7 %), *Shigella spp.* (6 %), ETEC (5 %), *Campylobacter spp.* (5 %) and *G. lamblia* (5 %). The etiological agents were not remarkably different from those of other African countries.

Diarrhoeagenic bacteria were detected in 112 cases (26 %). However, diarrhoeal cases with only bacteria were 75 cases (18 %). Diarrhoeal cases with only rotavirus were 57 (13 %), whereas cases with both rotavirus and enteropathogenic bacteria were detected in 12 (3 %) of the examined cases. A total of 61 (14 %) of the samples had parasites; out of these only 35 (8 %) con-tained parasites only as a potential cause of diarrhoea. There was one case (0.2%) from which bacteria, rotavirus and parasite were detected. The multiple infection with bacteria and parasite was the highest (25 cases, 6 %) among the cases of multiple infection. No enteric pathogens was found in 223 (52 %) of the 427 diarrhoeal cases.

Age distribution. The younger children had the higher incidence of diarrhoea. The mean age of diarrhoeal patients was 15.9 months. We compared the age of diarrhoeal patients with specific pathogens and that of patients without the specific pathogens. Statistical differences were seen in diarrhoea with *Campylobacter spp.* ($p = 0.0063$), *G. lamblia* ($p = 0.0006$) and rotavirus ($p < 0.0001$). *Campylobacter spp.* and rotavirus were found in the younger children. The mean values were 12.0 and 11.3 months of age, respectively. *G. lamblia* was found in relatively older children and their mean age was 22.4 months.

Seasonality of diarrhoea. The seasonal variation of diarrhoeal patients was found. Clear correlation between rainfall and the number of diarrhoeal patients was seen ($p = 0.0074$). In the long rainy seasons from April to June, the incidence of diarrhoea was very high. However, in the dry season, especially January and February of 1992, the number of diarrhoeal cases were drastically reduced.

In the rainy season, unhealthy situations like flooded floors were observed in these villages. The most serious problem was overflow of latrines. During the rainy season of 1994, a large outbreak of cholera and shigellosis occurred in this area and Mombasa, but this outbreak was not a part of this study. The problem of overflowing of latrines can be resolved, if a proper sewerage system is installed. However, the current economic situation is an obstacle.

Perceptions among People

When interviewed people know what the interviewer wants to hear. Since we were local people we soon had their confidence and were told the traditional explanations for diarrhoea.

- *Mavingani* i.e. when a lady shares a man with his daughter it is a bad omen and causes diarrhoea in the children or even in adults

- *Dege*, when a person sees you eating some food and you do not share it with him. He goes away without getting a piece, and you will end up with diarrhhoea.

- *Nyuni* is a certain bird and if the child is seen by that bird it will catch diarrhoea.

- *Katikiro,* i.e. adultery by a husband or a wife leads to diarrhoea in the children.

- *Jiko.* If one of the three stones in the fireplace is shifted or taken away it has a lot of cultural meaning. The lady in the house often keeps something under the stone and if it is removed for any reason it will be manifested in diarrhoea.

- *Nyumba*. If a man with two wives builds a new house, the first wife is to enter first. If the second wife or a niece goes first it will result in diarrhoea.

How do people manage diarrhoea? If the cause is as above, they have to seek advice from the medicine man or have to redo the thing, In the case of dege, it has to be done by smearing oil and water to get rid of the problem. In the case of *nyuni* they use some bones from the particular bird until the diarrhoea is gone.They also have they their own herbs they use either grind it and the child drinks it. *Mnuka uvundo* and fumigate the child in smoke.

Heat treatment of water

We designed an experiment to pasteurize drinking water and to find out at what temperature different bacteria are killed. We inserted different bacteria into the water and heated it to certain degrees using a thermosensor. At about 60 °C cholera vibrio was cleared, and other major causative bacteria were cleared at 70 °C. From April, 1995, we started to introduce this method to the villagers. A simple coloured slide was introduced indicating temperatures of 50, 60, and 70 °C with different colours. When villagers pasteurized their water the number of coliforms was markedly reduced.

People had to be motivated to use the method and their main concern was the fuel requirement. We illustrated the existence of bacteria by letting people compare the result of leaving some drops of boiled and un-boiled water on separate plates over night The next day the plate with untreated water had lots of growth, while the heated water left no trace. The common reaction was that this is what health staff mean when they talk about bacteria. And they felt there was a problem.

There are several advantages in this approach; 1) more effective and more introducible than chlorination and filtration, 2) lower risk of burning wounds and saving energy in comparison to boiling, and 3) easy recognition of 70 °C by the thermosensor.

We looked at some of the problems that people encounter in heat treatment of the water. About 50 % of the households have only two cooking pots, one for ugali porrige and one for vegetable stew. Some oil is left in the stew pot and it cannot be used immediately for heating drinking water. Crust of ugali in the pot needs time to losen which also make it difficult to use this pot. Plastic containers are common also in rural areas and cannot stand the heat of 70 °C.

Smoke problem when using green wood, When people are cooking for the evening and they proposed that they could use the ugali pot after it was cleaned. Scoop of coconut shell would be better to scoop water. The tea tastes smoke. We are still tracing effectiveness and sustainability of our trial.

Irregular visits to the villages indicated that areas served with tap water had less participation than those served by dams, ponds and wells. This is despite the fact that even tap water was found to be contaminated at the household level. This was due to the unhygienic carriage containers and household storage vessels.

COMMENTARY ON *THEME III.*

Re. Paper 6

Dr. R. Swiderski commented that the attitude of those who approach the community as scientists does not reveal the full picture as we have to consider the long term clean water situation and not just disease outbreaks. Although we do not have available a full picture of water use, we cannot assume arrogance in analysis, the communities are at present dealing with water without this knowledge.

The *Chairman* remarked that we are now seeing a wider environment of water use and we have to consider that the incidence of disease we would expect to see is not borne out by what we are observing; the incidence of disease is lower.

Dr. C. Walekhwa (Eldoret Municipal Council) noted that the incidence of disease is likely to be influenced by the prevalence of boiling water. He asked what percentage of consumers were drinking boiled water in this study. *Dr R. Swiderski* replied that the behaviour towards boiling water was erratic. People may some-times boil water, but not always, they may also return boiled water to a half full container of unboiled water. Roughly 25% of the population habitually boiled water, because they felt it helpful. But of course a lot of people who say they boil water do not actually do it.

Dr C. Walekhwa (Eldoret Municipal Council) mentioned that there are water kiosks supplying piped water in these areas. What interaction did the author observe between the use of piped sources and well sources? Dr R. Swiderski replied that entrepreneurs from outside are seen as being the vendors from water kiosks. Consumers felt the water was expensive and didn't like the taste. The kiosks however are probably more reliable which could account for the high incidence of purchase of piped water.

Ms A. Almedon (London School of Hygiene) noted that the results of this study concur with findings in Siaya district (see Almedon et al. 1994. Siaya Hygiene Evaluation Study. London School of Hygiene and Tropical Medicine), however observation alone is not enough, we learn more if we include interviews and this increases the sense of involvement of the consumers.

Mr M. Kong'ong'o (Moi University) felt that people say they boil water but it is doubtful if they give a true answer, observation shows a very low percentage of people actually boiling water. *Mr W.K. Sang* (KEMRI) mentioned that boiling water is very expensive, and if you decant hot water into the commonly used plastic containers the water smells of plastic. The Chairman remarked that you may not be offered water to drink from a glass because then you can see through it and observe colour and turbidity, instead you may be offered water to drink from a calabash. People often offer guests bottled or prepared drinks as they are seen as safer. There have been examples of people being offered local water to drink and it being suggested that if you wash it down with hot tea, the boiled drink will neutralise any effects of drinking the un boiled water.

Re Paper 7

Mr Haraka (Nakuru Municipal Council) pointed out that if there was an immediate impact of disease then people would boil water. In the event of disease outbreaks where the message to boil water is put across, people often continue the practice after the outbreak has subsided. In a case of a Typhoid outbreak, people did boil water, however the transmission route was traced to dried fish and ice in drinks. Once the people found the transmission was not water related they stopped boiling.

In his presentation *Mr Cairncross* referred to the response to an epidemic, which is often immediate and effective, yet there are often endemic cases of the same disease in that area which account for equivalent hardship for the population but over a longer period of time. Our response is more often geared toward epidemic rather than endemic disease. (The recent case of the Ebola virus outbreak in Zaire was mentioned). Just as many lives are lost through endemic disease as through epidemics. We have to consider hygiene in relation to health and endemic disease, with respect to water we should reconsider the plausibility of water as a transmission route and its relative threat compared with food borne routes. The seasonalities of disease often imply that transmission in relation to water may be over emphasised. It is however clear that effective hand washing is one measure which does reduced the incidence of disease whilst cutting across several transmission routes.

Dr R. Swiderski reported that in the areas studied near Eldoret, parents often said that the children became ill through eating contaminated food stuffs, not water. An example of how the local knowledge could be correct in the face of a different message being put across by the scientific community.

Mrs. T. Watts (Moi University) questioned the effect of washing our hands with soil rather than soap? *Mr Cairncross* responded by saying that there are many types of hand washing; in the domestic setting thorough hand washing has an important hygienic role. The effect of using soap is not just bactericidal, you actually have to wash the soap off your hands, hence you wash them thoroughly; using soil may lead to a similar thorough hand wash in order to get the soil off.

Mrs T. Watts (Moi University) also wondered what was the significance of water containers in the domestic environment for transmission. *Mr. S. Cairncross* (LSHTM) answered that there is contradictory evidence that containers may or may not be acting as reservoirs for pathogens. Obviously if they are kept clean a bio film cannot form on them, if they are narrow necked you cannot introduce dirty water scoops into them. But the risk to health may be far greater from other routes, water alone does not provide the substrates for pathogen reproduction, the issue of the storage container may be limited to only the survival of a limited number of pathogens and this leads us to the subject of relative infective doses and the hosts health status prior to infection. *Mr. T. A. Stenström* (Swedish Institute of Infectious Disease Control) mentioned the link between algal bloom and Cholera outbreaks and pointed to the possibility of algae in water containers offering a substrate for bacterial growth.

Dr.D. Bradley (LSHTM) commented that engineers are very preoccupied with water quality and this preoccupation has been transferred to the tropics. Yet there are a number of anthropological studies which begin to question the importance of quality, particularly when disease patterns are observed in groups of people whose access to water and behaviours are similar.

Mr. M. Woodhouse pointed out that even the well educated do not wash their hands when they should, and asked if hand washing was reasonable expectation. *Mr S. Cairncross* (LSHTM) responded by saying that the risks of not washing your hands in a developed country are much lower than for the tropics, in developed countries people are loosing that perception of risk as there are reduced consequences and are reducing their hand washing behaviour. *Ms. A. Almedon* (LSHTM) added that the importance lay in identifying the most critical times to wash hands and promote those in a community.

Re Paper 8

Ms. N. Balfour (AMREF) asked if *Mr Karama* determined what percentage of the community held traditional beliefs and what percentage held scientific beliefs. *Mr M. Karama* replied that no rigorous attempt was made to determine this difference, largely because it was known for example that a person who holds scientific beliefs may well turn to a traditional healer and vice versa. Overall it was felt that local people were more likely to take traditional action than scientific action.

Mr T.A. Stenström asked how long water samples were heated for at the various temperatures, *Mr Oundo* of KEMRI replied that the samples were only heated until the specific temperature was reached, and were not maintained at that temperature for a period of time.

PANEL DISCUSSION

ON PERCEIVED NEEDS OF SCIENTIFIC KNOWLEDGE AND INTERVENTION APPROACHES AMONG POLICY AND DECISION MAKERS

Chairman: Jan-Olof Drangert

Panel members were Mr. N. Waweru (Ministry of Health, Kenya), Mr E. Chege (Nairobi City Council), Mr A. A. Mohammed (Municipal Council of Malindi), and Mr R. Rotich (Eldoret Municipal Council).

In introducing the plenary session the chairman said that policy and decision makers have to strike a balance between common sense and among other things, scientific knowledge in the field of water supply and sanitation. For example, inspection by eye, smell and taste of river water quality provide the local people with a common sense assessment of water quality, while scientifically developed measurements of chemical constituents provide quantitative assessments. To decide upon what quantities of effluent are to be allowed into the river are not self evident.

One example of policy statement is the WHO guidelines recommending zero faecal coliform bacteria in drinking water. This is a too easy way out in real life situations. It is like recommending that no cars and lorries are allowed in our streets in order to prevent accidents.

The panel members gave their ideas on the synthesis of the presentations and raised additional important issues.

Mr N. Waweru said that suitable technology remains an issue. We have permeable soils and high water tables, we promote the latrine but as yet have still to identify a suitable design. Also land pressure in periurban areas means there is little land on which to build latrines once the old ones are full.

Mr S. Cairncross (LSHTM) pointed out a number of composting latrine designs which are performing well under these circumstances. He also described types of latrine emptying vehicles, such as designed by the Building Research Establishment (BREVAC) in UK which is vehicle mounted. Although the vehicle works well, it has a high maintenance cost and the truck mounting means it cannot reach all of the latrines in a densely populated area. An alternative called the MAPET which is manually operated and mounted on a hand cart has been used in Dar es Salaam. Several areas where manual emptying by a digger scooping or shovelling the pit clean were mentioned by the participants.

Mr A. A. Mohammed began by noting the issue of the safety requirements in manual pit latrine emptying. Mr Mohammed then gave an example from his many years work in Eldoret. In the 1970's when the periurban settlement in Huruma was beginning the council had to oppose the development under the existing building codes. 150 people were prosecuted and moves were made to begin demolition of the offending buildings. The owners lodged an appeal in the high court and demolition work was suspended. In the meanwhile more people were building houses. The case stood still for a year. In that time the settlers formed two land companies.The council agreed to meet these companies to explain the purpose of the relevant legislation. The outcome of the meetings was that the companies agreed to survey the land and prepare a plan for its development, which would be subject to council approval. The plans were approved and title deeds for the land issued. In this case the law would only create more enmity than dealing directly with the people. The council recognised that other areas of the town would soon be undeı pressure for settlement and began to follow similar procedures when dealing with new settlers.

In his present work in Malindi on the Kenya coast Mr Mohammed described a range of primary health care activities in the periurban areas which included solid waste clean ups and the provision of piped water supplies. Mr Mohammed raised two issues, firstly the question of sustainability of interventions and secondly what constitutes satisfactory treatment of well water.

Mrs T. Watts responded by mentioning references in the literature to simple domestic sand filters and the production of sodium hypochlorite using solar energy. In *Mr. Cairncross's* opinion neither of these techniques were easy to use and their usefulness was dubious.

In a open discussion amongst the participants it became clear that the issue of sustainability would require considerable time to debate, amongst the points raised was the importance of responding to felt needs and participation of the people particularly in designing solutions.

Mr E. Chege raised three issues for consideration. The main trunk sewers of Nairobi are not working. This results in gross faecal pollution as sewage overflows manhole covers, and emerges at ground level, eventually reaching the rivers draining Nairobi. The problem remedial work presents to a large city is considerable.

Secondly the legislative route to the abatement of public health nuisances does not work effectively. It is current practise to address the problem to the individual responsible, but a whole regime of delay tactics has emerged via the use of injunctions which can take up to a year to resolve. Meanwhile the nuisance remains unabated.

Thirdly Mr Chege pointed to the lack of coordination amongst the NGOs in the provision of drainage, latrines and refuse collection for the residents of the periurban areas.

Dr C. Walekhwa (Eldoret Municipal Council) said that the Nairobi city task force had been empowered to deal with many of these issues and wondered as a body of professionals to oversee other professionals how it had succeeded or failed. *Mr Chege* responded by saying that although the task force could prioritize its work it found the same conclusions for the failure of systems as the Nairobi City Commission had done before it. In respect to the issue of blocked sewers, numerous reasons can be ascribed however. The real problem becomes a question of the commitment to keeping them free flowing.

Mr R. Rotich raised four issues of concern; the need to promote the control of vectors of disease, the fact that there were no primary health care programmes in the Eldoret periurban areas, the issue of waste recycling (in particular the increasing presence of non biodegradable plastics) and finally the presence and effects of vandalism on public utilities such as latrines.

Mr M. Woodhouse responded that even though biodegradable technology exists the issue should be addressed to consumers and marketing strategists to reduce the demand for such potential waste.

Mr J.O. Drangert said that in Europe it is estimated that 90% of what has been produced has yet to be effectively disposed of. Information about solutions exist but are not disseminated. Overall people's behaviour appears to be critical and should be addressed.

THEME IV : DUG LATRINES AND GROUNDWATER CONTAMINATION

Paper 9. TRACING BACTERIA FROM THE LATRINE TO GROUNDWATER IN DUG WELLS

Thor Axel Stenström, Swedish Institute for Infectious Disease Control, S-105 21 Stockholm, Sweden

Groundwater and Factors Governing Contamination

Groundwater has been used from the earliest time as a prime source for household use, for livestock as well as for agricultural purposes. Its existence and use also has deep roots in many traditional religions, from the old Asa-beliefs in the Scandinavian countries to traditional beliefs in East Asia. The holy and healing spirit of springs have been part of many cultures.

Groundwater is also a dominant portion of the existing freshwater on earth, which is often overlooked. If polar ice-caps and glaciers are excluded groundwater constitutes about 95% of the world's freshwater resources. Lakes, swamps, reservoirs and rivers only account for about 3.5% and soil moisture an additional 1.5% of the freshwater.

Groundwater is present in thc saturated zone, where water fills most of the void space between soil particles, cracks and fractures in rock formations. Groundwater is always on the move, but it often flows very slowly.

Groundwater is vulnerable to chemical pollution and the water tend to remain polluted since the horizontal filtration and adsorption is minimal. The natural rate of clearing may take years or even decades rendering the groundwater unfit for use for long periods of time.

Groundwater may, however, flow much faster. These underground rivers in "aquifers" of high permeability are often preferred as groundwater sources since they give high yields and are less prone to dry up. In sloping terrain hydrological gradients are formed and more rapid flow will occur. Just like a river above the ground, groundwater will flow downhill

Groundwater will be formed partly from influxes from rivers, lakes and swamps, but mainly through infiltration of rainwater through the soil. The rate of infiltration, which is a function of the size, form and distribution of the soil particles will govern how far down pollutants will travel into the ground with the infiltrating water. The deeper the unsaturated zone is (the depth from the ground surface to the groundwater table) the lower the probability of groundwater contamination. In the unsaturated zone the microorganisms in the water are travelling a far greater distance than in the saturated zone, since they have to follow the thin water layer/film on the surface of the soil particles, while in the saturated zone they are transported with the flow in the water-filled pores and fractures. This means that in areas where the groundwater table is deep below the surface there is a lower probability of contamination from microorganisms.

The size of the organism as well as factors related to the possibilities for adsorption of organisms onto soil particles, govern the spread of organisms in the groundwater. Parasitic eggs, cysts and oocysts, although small, are relatively large from a microbial perspective and are less prone to be transported through the soil, while bacteria and viruses, partly due to their smaller size have a much higher possibility to be transported.

When digging pit latrines for faecal disposal we are reducing the distance that the organisms have to travel in order to reach the groundwater. We have also reduced the natural filtration effect of some of the unsaturated zone. If in addition to that the horizontal permeability is high (i.e. where there is a hydrological gradient) there is a high likelihood for an organism to spread from the latrine to nearby wells. The situation will be worse seasonally with a high or fluctuating water table.

Different pathogenic microorganisms have varying ability to survive in water and soil. This ability is partly species and strain specific but is also due to prevailing environmental conditions. The conditions existing in groundwater will often favour prolonged survival.

Indicator Organisms of Microbial Pollution

Different microorganisms occurring in faecal material are used to estimate hazards of microbial pollution of water. These are normally not disease-causing organisms and occur naturally in high numbers in faecal material. If found in water, their presence indicates that faecal pollution has occurred and their numbers indicate the extent of such pollution. Much used indicator organisms are for example *faecal coliforms* and *faecal streptococci.* Their presence indicates that directly pathogenic bacteria, viruses or parasites may also be present.

A dug well is prone to be potentially polluted from many different sources, especially if uncovered and not protected. Pollution may occur due to human practices during drawing the water, or due to accidental contamination. Unprotected dug wells may also be polluted due to water handling and surface water intrusion directly in the well or in the nearby area around the well and from groundwater contamination. Using indicator microorganisms we can not differentiate between the possible pollution sources. For such a differentiation of routes the differernt waters need to be tagged individually and the water in question traced from its source.

Tracers of Water Flow

Several methods have been applied to trace a water flow and to estimate the dilution of organisms during transport. These methods include the use of salts like sodium or lithium, radioactive isotopes, fluorescent or non-fluorescent dyes, and different groups of microorganisms. Many of these methods are less applicable in relation to existing water sources for human use. Furthermore, investigations have mainly been done in developed countries.

Problems have occurred when trying to apply chemical tracers to estimate microbial transport rates. Over short distances the transportation rates seem to be similar, however experiments using non-reactive tracers have shown a more rapid microbial transport. Fluorescent dyes like fluorescein and rhodamine do adsorb to a greater extent to soil particles. A similar non-correlation exist for positively and negatively charged dyes.

Using microorganisms as direct groundwater tracers has the advantage of a high sensitivity as well as directly mimicking the organism under investigation. Yeasts have been used but are less effective due to their size, they may however be useful in estimating transport rates of some parasites. Different bacteria have been used with special physiological or morphological characteristics, but their use is limited especially if the water is used for human consumption.

Therefore, bacteriophages (viruses only attacking bacteria) may be a better choice.

They are of the size of enteric viruses, thereby mimicking the risk for viral spread but are harmless to humans, animals and plants. Strain selection of phages results in extended survival times, easy propagation in high concentrations, and no sophisticated instrumentation is needed to detect them. They can be assayed from water samples within one working day. They are in many ways ideal as markers for different transmission routes. Their use as tracers is increasing, McNabb (1979) used two different phages to trace contamination from a land application site and Skilton & Wheeler (1988) used different phages for groundwater investigations in England. Some investigations have also been made in Malawi.

Tracer Investigations of Groundwater Contamination in Eldoret.

In the present study the aim was to explore possible groundwater contamination from dug pit latrines in two periurban settings in Kenya, where many dug wells occur and latrines are in contact with groundwater, at least seasonally. Along the East African coast other studies have shown leakage and heavy groundwater contamination through the coral-based ground.

The area under investigation in Eldoret municipal was, however, characterised by a phonolite-murram ground, with low vertical water penetration. The study was performed in one area, Huruma, with an obvious hydrological gradient and in one without, Langas.

The specific objective was to evaluate whether the present distances between latrines and wells under these conditions resulted in contamination of the magnitude of an infective dose. A further evaluation of this route of transmission in relation to other possible routes is underway.

A special strain of phages was used in the tracer experiments. A solution with a known amount of phages was inserted into the faecal material in different pit latrines (about 6-10 metres deep) with the aid of a rigid plastic pipe. Water was then added through the pipe. The water in adjacent wells were sampled daily or at intervals and analyzed for the presence of the phages. Control experiments were performed prior to the application showing no background presence of the special strain of phages.

Phages were recovered in wells in 5 of the 6 tested latrine areas. In several locations it was just a matter of one or a few days before wells in the neighbourhood of the latrine showed positive for the tracer phages. However not all wells tested positive, and neither did some wells which were the closest to the experimental pit latrines. Naturally this depends on the direction of the groundwater flow. In one of the studies in Huruma with a steep hydrological gradient, an abandoned well which was located 130 metres from the pit latrine became positive within one day. This was the longest and most rapid transportation encountered. Normally the wells which turned out positive were situated from a few meters up to 40 meters from the pit.

This was not just the result in Huruma, but also in Langas, with its flat ground surface. Here, underground transportation of phages reached wells up to 40 meters away within a week (further distances were not recorded as no wells were selected beyond this distance). Another situation should also be recognised in which groundwater flow is not uniformly radial but follows discrete paths which are not immediately obvious, but dictated by local geology. In one study in Huruma all wells downstream of the pit latrine turned out negative. However, when the study was repeated several months later and the phages introduced into a different pit, several of these wells turned out positive.

Discussion.

Several questions and recommendations could be put forward based on this study.

Is there a safe distance between a latrine and a well. The present rule of thumb requiring a safety distance of 10 to 30 meters between a latrine and a well, is not universally acceptable. The safe distance depends upon the prevailing on site soil conditions. In the present study it was shown that a spread of over 40 m was possible in a matter of days in areas without an apparent hydrological gradient. Under the prevailing situation of congested settlement in periurban areas, few alternatives to pit latrines and shallow wells exist. Despite the fact that there are some piped water and a few plot connections in Huruma which the tenants claimed that they use, observation proved that they rely on the wells. In Langas the water kiosks are not an effective alternative source as the water is costly and far from the home.

More thorough pre-investigation needs to be carried out before a pit latrine programme starts and must take into account the local hydrogeological conditions. If pit latrines are used for faecal disposal and residents rely on wells by necessity or habit other alternatives for disposal should be discussed. Composting toilets may be an alternative.

Should the depth of a pit latrines be restricted ? The answer depends on the prevailing conditions. It is necessary to take the groundwater level into account. There is clear evidence that microorganisms spread is much enhanced when they are introduced directly into the groundwater. A safety limit of at least 1 metre but preferably 3 metres should exist between the bottom of the pit and the groundwater table. The limit must account for seasonal fluctuations in the groundwater table and a pit should never be excavated deeper than one metre above the highest groundwater immunosuppression. Other agents or transmission table. High soil porosity and the presence of fractured rock beneath the pit would necessitate more stringent safety limits. It is very necessary to take into account the polluting effects of pit latrines in urban areas where deep groundwater is abstracted and distributed as tap water. Many examples from Europe and the United States show that on-site sanitation units similar to pit latrines have caused waterborne outbreaks of diahorreal disease due to seepage through fractured rocks.

However, if an area is assigned for pit latrines and no wells exist and drinking water is provided by other uninterrupted services, then the above stated conditions do not have to apply.

Alternatives for pit latrines do exist in different types of composting toilet systems, built-up soil infiltration units or specially designated infiltration areas. A waterborne sewerage system is therefore not the only choice.

Will low amounts of pathogenic micro-organisms seeping through the ground pose a health risk for the exposed population ? This question may have many answers. In areas like Langas and Huruma, with unprotected wells without handpumps, and where the water is more grossly contaminated by other routes, (such as surface drainage water intrusion and poor water drawing practices), a low level of contamination through the groundwater will probably play a minimal or marginal role. However, an important outcome from this study has been the knowledge that the distances which viable microorganisms are transported underground can in fact be far greater than previously expected. This is important where upgrading of wells to improve protection or the introduction of boreholes are being considered. In such situations the other possible routes of transmission are eliminated and the transmission through groundwater will be the primary one in relation to the water at source. In these instances the groundwater pathogen transmission may become the route of major concern.

Many more answers are possible to this question. The transport through the groundwater from the pit latrines will most certainly eliminate the spread of most pathogenic parasites and protozoa (a question mark may be raised about the oocysts of *Cryptosporidium*). It would also reduce the amounts of many bacteria below the infective dose. However taking into account the number of tracers applied, the amounts reoccurring in the wells will be sufficient to cause infections by

Shigella, Campylobacter, S. typhi and haemorrhagic *E. coli*, but will probably exclude toxigenic *E. coli*, other *Salmonella* infections, *Vibrio cholera* and so on. These small amounts are also sufficient to cause infections by all of the enteric viruses. The discussion is partly hypothetical since we have only taken one pit into account for the spread to a specified water well. Pollution from several different latrines is likely to take place simultaneously. The dilution rates are also likely to vary from place to place and also seasonally.

The recharge of surface water to the groundwater seems to be low judged from the large amounts of standing water and run-off after heavy rains. However this apparent low infiltration may be a miss judgement, and areas of high infiltration exist in between.

This leads us further to the effects of previous exposure and development of immunity against these organisms. For people without immunity the lower doses are sufficient to cause disease. This has been proven in another situation in Sweden, where a similar amount of *Shigella*, as demonstrated by the phage tracing, was spread through groundwater from an on-site soil infiltrating unit used by a Shigella-infected person to a well about 50 metres away. This gave rise to another case of the same strain of Shigella. No personal or other contact existed between these two persons. However in an environment like Huruma and Langas, there is a high likelihood of immunity or partial immunity for many of these pathogenic agents. Exceptions would be for example the sick or non breastfed babies, the malnourished and people with routes may however be more important for these groups as well.

Exceptions however occur within a society if there exist groups that are separated from each other for ethnic, cultural, religious or other reasons.

The question can also be raised wether exposure to low levels of pathogens is beneficial for society by continuously boosting the immune response for some of these potential pathogens!

It can therefore be argued that in settings like Huruma and Langas with the seasonal effects of standing water, highly faecally polluted storm and drainage water, and where individuals use other more polluted drinking water sources than their wells; individual water related behaviours may pose a greater risk of infection than those arising from the "background contamination" of the groundwater.

Is faecal contamination of groundwater a major concern in relation to disease transmission from latrines ? Disease transmission from latrines to groundwater may be less significant than direct transmission to people from faecal material from the latrine floor and ground close to the latrine. Washing of hands does not exclude the spread of faecal material by bare feet. Children play around the latrine and will evidently get their share of organisms. Direct or secondary transmission will therefore take place.

Several subgroups of people may not use the latrine, for example pregnant women, children under 5, the old, the sick, or people that disapprove of the presence of faecal material on the latrine slab. Taken together this may be a substantial part of the population. The faecal material from these people will either be left on the ground (partly scavenged by dogs, pigs, ducks and chickens) or appear in the gutters and will form a substantial possibility for secondary transmission.

In addition secondary transmission may occur from flies in the latrines, with cockroaches, other insects and rats. It may be interesting to see, if the distance from the latrine or ground areas of faecal contamination may be of higher importance in relation to diahorreal cases. Here, dwellings close to a latrine may be more vulnerable in relation to secondary transmission with insects and rats.

Secondary transmission may also occur from domestic animals. The common practise of keeping chickens and ducks within the houses may lead to such transmission.

Influence of person to person spread in relation to contaminated water. A lot of people living close together will face a higher likelihood of getting into close contacts with each other. Even if erratic hygiene practices occur (handwashing; washing in the evenings), this will not exclude the multitude of contacts during the day, and the in-between-contact with pathogens. Children are in constant contact both with each other and soiled ground or items, adults are also in constant contact, shaking and holding hands and so on. People go to work in other localities, relatives and visitors are entering, giving rise to a constant flux of microorganisms. This will naturally affect the possibilities of disease transmission to a great extent and affect vulnerable individuals. The number of influxes and outfluxes, contacts with other groups as well as individuals may be of much higher importance than a low degree of pollution through water.

Different water sources may however also play different roles in this import and export of microorganisms to a defined groups of people, giving the highest likelihood of transmission through severely polluted river and surface water than marginally affected groundwater.

References

Skilton, H.E, and D. Wheeler. 1988. Bacteriophage trace experiments in groundwater. *J. Appl. Bact.* 65, pp. 387-395.

Paper 10. GROUNDWATER CONTAMINATION FROM PIT LATRINES IN PERIURBAN DAR ES SALAAM

Shaaban Mgana
Ardhi Institute, P.O. Box 35176, Dar es Salaam

Introduction

A small fraction of the population is serviced by sewerage systems, while more than 90% of the Tanzanians use "on-site waste disposal". Pit latrines account for 90% of these and the remaining 10% are septic tanks. This was the situation in 1985 and a similar scenario might exist for quite some time.

In 1971, the Tanzanian Government declared a policy of supplying clean and potable water to every person by the year 1991. That declaration has not been realized. The Government's goal has now been pushed forward to the year 2002. However, the mechanisms of achieving this goal have not changed. Technology mix is the plan of action; whereby surface water and groundwater will be developed based on least cost criteria and availability in the respective locations.

It is essential to make sure that the commonly practised on-site waste disposal systems do not pollute the groundwater, since this resource will remain a crucial water source and is often used without treatment. Even in those areas which are wholly provided with piped water supplies, there are frequent breakdowns which have forced people to turn to hand-dug wells or standby shallow wells whose bacteriological quality is usually taken for granted.

It was against this background that ground water pollution research by Ardhi Institute began in the early 1980s. The research was conceived to scientifically unveil the interaction between water supply and existing on-site waste disposal systems at the local level, particularly in unplanned periurban settlements. The periurban area, Majumbasita and some neighbourhoods outside Dar es Salaam were chosen.

Objectives

The overall objective of the research was to investigate the contamination of groundwater and of tap water supplied by the municipal in areas where on-site waste disposal systems are used, in order to develop guidelines for the siting of these facilities to reduce the risk of water contamination. The specific objectives of the research were:

- ◊ Preparation of an environmental map showing all existing on-site waste disposal systems and dug wells/boreholes as well as the location of municipal water distribution mains and undertake a user utilization survey of these facilities in the identified periurban areas.
- ◊ Hydrogeological studies of the groundwater flow regimes and determination of the risk of bacteriological contamination of traditional dug wells from on-site waste disposal systems.
- ◊ Installation of six demonstration boreholes equipped with Mark III hand pumps to determine the potential of deeper zones of the aquifer to provide potable groundwater.
- ◊ Evaluation of user satisfaction and bacteriological water quality of water produced by the demonstration boreholes compared to the traditional wells in the study areas.
- ◊ Development of guidelines for the siting, design and construction of wells/boreholes and on-site waste disposal systems with reference to specific hydrogeological parameters.

Implementation of the study

The methodology included:

Literature review. The researchers consulted a number of sources within and outside Africa on groundwater pollution. A considerable amount of data is acknowledged on environmental pollution, pathogen die off and the movement of anions, cations and pathogens in groundwater. The information found in the literature review is mainly applicable to temperate climates of Europe and North America where the investigations were carried out. The literature search showed that climate influences the survival of microorganisms together with a complexity of environmental factors which do differ greatly from one geographical location to another. As a result the observations made in Europe and North America cannot be used outright as the basis for making a reliable environmental pollution assessment in the tropics without first testing their applica bility locally. This research provided such an opportunity for making local assessment of ground water pollution as caused by on-site waste disposal facilities.

Environmental base map. The environmental base map below was prepared using land surveying techniques. The map shows locations of housing units, traditional sources of water supply (dug wells, springs and streams) and all on-site waste disposal systems such as pit latrines and septic tank systems. Any unusual topographical features such as gullies and swampy areas are shown. It took four months instead of one month as planned for the eight surveyors to complete this work.

A total of 1292 built structures and 687 on-site waste disposal systems were plotted. Out of these on-site waste disposal systems 581 (85%) are pit latrines and 106 (15%) are septic tank systems. In the area there are 37 dug wells, mostly used temporarily used for various purposes and one perennial spring.

Map of the study area including housesand water sources (W)

Installation of piezometers. Piezometer installation in the study area has been the most tedious work. It took one month for seven people to install 25 piezometers using hand augers. The identification of sites for installing piezo-meters was done by the Principal Investigator in collaboration with a team of consultants from Canada, named Water and Earth Science Associates Ltd. The consultants also provided the expertise of installing the piezometers into the ground. Their contribution to this project is very much appreciated.

The project area has been divided into 35 sub-areas for hydrogeological study convenience. A total of 25 boreholes have been installed in the project area. It has been found that the study area has a shallow unconfined aquifer separated by a thin clay layer from a deeper aquifer. In some places the unconfined aquifer is located at a depth of about 1.5 to 2 metres below ground level. The deep aquifer is located at a depth of about 3.5 m., It was observed that the direction of flow of the groundwater for the shallow piezometers is at approximately 45^0 to the direction of groundwater flow for the deeper piezometers. Most likely the shallow groundwater flow is responsible for the seasonal springs which happen to be located in the direction of this shallow flow regime. On the other hand, the dependable and perennial spring could be originating from the deep ground water flow regime.

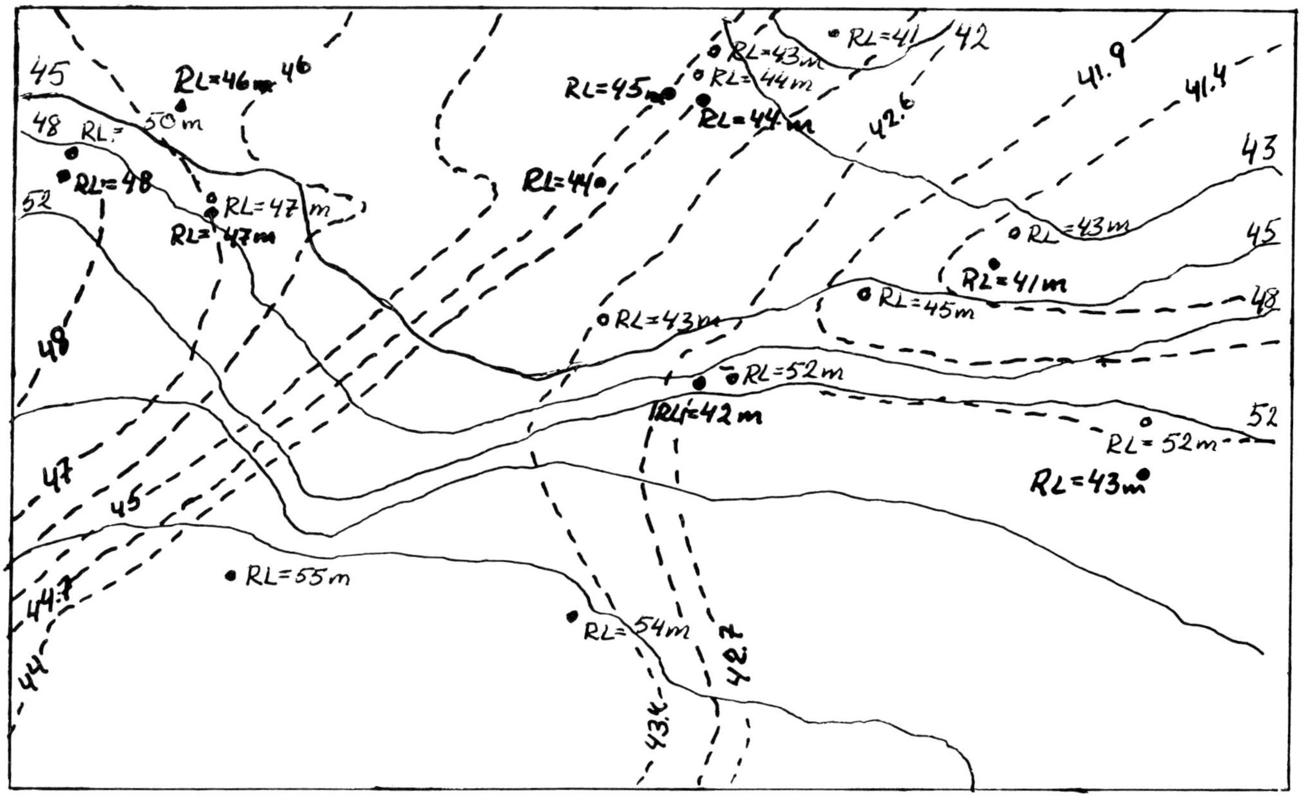

Fig. Groundwater contours for shallow (⁀) and deep (- - -) groundwater boreholes

Laboratory analyses. The analysis of soil samples taken during the augering exercise were done by the Soil Mechanics Laboratory of the Ministry of Water, Energy and Minerals. Whereas the analysis of water samples drawn from the piezometers and municipal water distribution mains in the project area was done in the laboratories of Ardhi Institute using the project's equipment.

User utilization survey. The aim of conducting a User Utilization Survey of the project area was to conduct an on-site field inspection of the sanitation situation. It involved inspecting the sanitation facilities to get comments from the users of these facilities. Random numbers were assigned to all built up plots and 500 houses were sampled and earmarked for interview, constituting about 38 per cent of all the built up plots as found in the environmental map.

Limitations. Going through the activities which were to be accomplished during the research period one would find that not all of them have been done. The installation and testing of six demonstration boreholes equipped with Unimade Mark III hand pumps and subsequent evaluation of the user satisfaction and bacteriological water quality of water compared to the traditional wells in the study area was not done.

Research Findings

The research findings are summarised as:

(a) About 60% of the people interviewed in Majumbasita said they found no problem with pit latrines. Even bad smell from pit latrines did not seem to be a problem to them.

(b) The public did not give pit latrines and superstructures a priority comparable to their houses which are normally better constructed. It was found that about 48 % of the latrines sampled had no roofs.

(c) The residents prefer big pit latrines with average size of about 13.5 m^3.

(d) Every inspected pit latrine had a small open tin/plastic container. The average volume of the containers was about 1.1 litres. Presumably their main function is to hold water for ablution purposes. If we assume three latrine visits per day per person from a house of 9 people using the same pit latrine, then this would yield a water load of about 30 litres per day. Normally 30 litres can easily soak away in the pit under reasonable ground conditions, particularly during dry seasons when the groundwater lowers. However during the rainy season ground-water contamination is likely to occur.

(e) The presence of water containers in every latrine visited suggests that introduction of pour flush latrines (which are hygienically superior to the traditional pit latrines) could be well received, if the public is educated and backed up by legislation.

(f) Preferences on methods of emptying the pit latrines are: abandoning 21 %, manual emptying using traditional technique (frog man) 28 %, using vacuum tankers 45 %, and undecided respondents 6 %.

(g) However, some 92 % of the sampled pit latrines have never been emptied.

(h) The average age of the pit latrines sampled was 9.6 years. However the oldest pit latrine was 31 years and had been emptied twice. The oldest pit latrine sampled that has never been emptied was 27 years.

(i) Almost all pit latrines sampled in the study area were raised above ground level. The raised portion of the pit latrine gives the pit content some extra room for rising and lowering in phase with the groundwater level depending on the season. Without the provision of this extra space many pit latrines would be overflowing during the rainy season.

(j) About 78 % of the sampled pit latrines have been constructed by hired labour. This means that any proposals on the legislation and technology with regard to siting, design, construction and maintenance of pit latrines and septic tanks should be disseminated to local masons. The legislation and technology should also be included in the curriculum of vocational training colleges. Despite the importance of on-site waste disposal systems, there is no legal requirement for involving/ consulting the relevant experts for the construction of pit latrines and septic tank systems.

(k) The shallow groundwater (unconfined aquifer) level fluctuates between less than a meter to 7 m below the ground level.

(l) In most of the study area the top/ shallow groundwater table is only 3 m below ground level and is faecally contaminated.

(m) Most of the domestic water points sampled from the piped water supply were also faecally contaminated.

(n) The data show that about 55 % of the people surveyed do not boil drinking water, particularly if the source is the piped water supply.

(o) The literature review revealed that legislation in Tanzania governing the siting, design and construction of pit latrines and septic tanks is inadequate and out of date. However there is adequate legislation (Public Health and Sewerage - Rules 1955) which governs the sewerage system despite the fact that very few people, about 10% of Tanzanians use them.

(p) There is no legislation or rules on specifications and detailed requirements for the construction of latrines and septic tank systems.

Discussion

Conditions in Majumbasita can be summarized as potentially unhealthy for the residents of the area. The situation has gone this far due to lack of a reasonable sanitation infrastructure in the area. It is the responsibility of the government to provide services like sewerage systems and reliable water supplies.

It has been a common practice for many governments to finance sewerage systems which serve fewer people and to provide little or no funds for low cost sanitation options with comparable if not better sanitation advantages.

In the case of inability of the government to provide such essential public services then automatically the affected people resort to solving their problems within their socio-economic capability. Such sanitation facilities are normally below the required standards. But in this case it has been observed that there are no legislatively prescribed standards to be followed with regard to on-site waste disposal systems (i.e. pit latrines and septic tank). That means even if the public wanted to construct affordable sanitation infrastructures there are no standards to conform with. The outcome of this is the unguided mushrooming of poorly sited, designed and built systems which pollute the groundwater resource.

Proposals

Based on the existing situation, the following proposals are made with the view of alleviating the sanitation problems facing many peri-urban areas of similar nature to the study case. The proposals are presented in a mixed form.

◊ Governments should enact detailed appropriate legislations on specifications and detailed requirements for on-site waste disposal systems with regard to siting, design, construction and maintenance.

◊ Governments should provide parallel assistance to the public for improving the pit latrines and septic tanks. On-site waste disposal systems cost US$ 80 - 240 for a pit latrine which could serve about 9 people. That is about US$ 9 - 27 per person as compared to sewerage systems which cost between US$ 350 - 500 per person.

◊ The existing sanitation options of a ventilated-improved latrine or pour flush latrine versus a conventional sewered system provide too narrow a choice between two extremes. Both systems are not sustainable due to some limitations.

Sustainable development requires development and implementation of methods and measures that lead to minimum consumption, lower production of wastes and maximum recycling. In this regard this paper calls for the development of such a system, preferably a community on-site waste disposal system.

Conclusion

The major lesson learned from this research is that it is very difficult to prevent groundwater contamination particularly in an unconfined aquifer in human settlements which employ pit latrines and septic tank systems at every housing unit. In this regard the pit latrines / septic tank systems become so numerous for the variable hydrogeological environment. It is too expensive both in terms of finances and technological effort to ascertain the hydrogeological environment at the individual house level, such that it is not possible to accurately advise each householder where they should locate their on site sanitation facility.

This realization leads to a fundamental concept that the best approach in minimizing groundwater contamination in periurban human settlements is through the provision of community on-site waste disposal systems which could be well managed. Otherwise the groundwater and the health of the people will be highly vulnerable.

The approximate locations of shallow piezometers are presented in map No.4, which also shows the extrapolated ground water contours. The layout of the deeper piezometers and the associated ground water contours are shown in map No.5. It should be noted that map Nos. 4 and 5 are approximations which are yet to be confirmed by actual survey and levelling that will determine the exact spatial locations of the piezometers and their respective ground level altitudes.

Paper 11. LOCAL PERCEPTIONS OF GROUNDWATER CONTAMINATION

Melvin Woodhouse
P.O. Box 9200, Dar es Salaam, Tanzania.

Introduction

In this session of our meeting we address the issue of groundwater contamination and in particular the interaction with on site sanitation. In 1908 the problem had already been minuted by the Mombasa Water Supply Committee in Kenya; which was eager to press its case due to an increasing population stressing the existing wells on the island (1). By 1917 a piped supply had been built for the island which enabled the expansion of the harbour facilities and gave subsequent financial return. Today we find ourselves debating a similar issue but in the absence of the pipeline option, we are seemingly trying to apply intermediate technology which is potentially incompatible yet the best available given the current resources.

I shall present work which has elicited and built on the local perception of groundwater contamination. Increasing local knowledge is seen as a valuable step in improving both sanitary and behavioral safeguards of water quality, a way of lessening the negative effects of the technologies we are employing. Yet practising good sanitary behaviour and making the long term use of local perception of groundwater quality purposeful remain elusive.

For the purposes of our discussions I hope that we can leave our definitions of what constitutes "local" and "perception" as negotiable.

Local Perception of Groundwater Quality

Practical examples of developing and incorporating local perception skills around groundwater quality issues are numerous. The opportunities such work opens for all players in the development sector are ever widening and challenging. I have adopted to describe some of this work by looking at what is immediately perceived, what can be perceived over the longer term and those other aspects which lie outside this split.

Acute perception. From the water scientists' perspective organoleptic parameters which are readily observed by the consumer include colour, turbidity, taste and odour and are "usually having little health significance" (2). Although colour and turbidity can be readily quantified, the perception of taste and odour varies greatly from one person to the next. What we determine as acceptable depends upon what we are used to and our body chemistry, and there are always complaints. Generally, the taste buds in the oral cavity specifically detect inorganic compounds of metals such as magnesium, calcium, sodium, copper, iron and zinc (3), all of which are apt to appear in groundwater. Consumers often describe these tastes as metallic, salty or chalky, and it is not uncommon for people to mix water from two sources to improve the taste. A common problem with groundwater is its salinity and average salinities of East African shallow groundwaters are some three times higher than the EC guide level of 400 mS/cm. Even so in some communities the people do not object to drinking water which is moderately saline, but there is no clear cut off point (4).

Even when consumers inform us that the water is too salty, removal of salinity is costly and it is therefore more pragmatic to identify salt free sites for development.

Table 1. Consumer objections to salinity: data from Kibwezi division Kenya

Salinity (mS Well /cm)	Consumers Objecting to Taste (%)
483	17
734	0
781	60
1349	0
1458	18
1880	63
1990	0
2800	29
6980	91

Source: AMREF, 1989

Turbidity is less of a problem in a properly protected groundwater source, but sometimes a high iron content may impart a brown colour or manganese a black colour in the form of deposits or stains where water spills from a pump onto the well head.

Of far greater immediate significance is bacteriological contamination. In addition to the work of Dr Mgana and Dr Stenström presented in this conference a growing body of evidence is showing that a pure groundwater source is commonly highly contaminated in the process of drawing, transporting and storing water from point sources such as wells (5 & 6). In the latter reference 87% of water samples taken from the homes of people using a well containing zero faecal coliforms were found to contain 10 or more faecal coliforms per 100 ml. In such surveys it is not enough to sample the water by aseptic technique at the pump. The picture revealed when sampling from a glass of water offered at the home of the consumer is very different and demands action in the domestic domain rather than just at the public water point.

A KAP-study followed up the previous done work in Kibwezi, a rural area on the coast (7). Musomi found that 93 % of well water samples contained less than 10 faecal coliforms per 100 ml and that 94% of home samples contained more than 10 per 100 ml. Although 73% of homes had pit latrines, the high level of contamination of domestic water was not related to the presence or absence of a latrine and neither was it related to distance to the latrine. Significant correlation was found, however, between low levels of domestic water contamination and homesteads in higher socio-economic groupings and homesteads whose heads had secondary or adult education. It should be mentioned that most people in Kibwezi collect water in plastic jerry cans which have small openings, the same container serves to transport and store water in the home. Dirty and stained jerry cans did seem to support higher faecal coliform numbers but this was not proven with statistical significance.

Musomi's work prompted interesting debate at the project. It appeared that wells can certainly provide good quality water, but if people do not practise what they know to be hygienic behaviour they should not expect this water to remain of good quality as they use it. Musomi did however point out that even though there was good knowledge of hygienic practises such as protecting water storage containers, knowledge about the causes of contamination of well water was less complete. Of 258 respondents, 70 % recognised cracks in the well head, 49 % defecation near a well, 32 % poor drainage around the well and 16 % via seepage from latrines as causes of well water contamination. But it is rare to find latrines within the recharge areas of shallow wells of Kibwezi.

We wondered how much "good behaviour" could be expected from people and an impromptu survey at the African Medical Research Foundation (AMREF) headquarters in Nairobi revealed that only 60 % of the people using the gents toilet on the directors floor washed their hands afterwards. We were given the idea for this survey by a colleague who had done the same at the Liverpool School of Hygiene and Tropical medicine where they were found to be 70% hand washers.

Given the difficult conditions under which people live in Kibwezi, it is not surprising that only 14% of people there said they washed their hands after using the latrine. In Kibwezi there is no water on tap at the latrine, water is scarce (average consumption 11 lcd) and so you would have to use "carried" water from the domestic storage place if you wanted to wash your hands.

Others in the project argued that we must improve our engineering approach to make it easier for people to practise good hygiene. Some steps were taken in this direction. A programme to assist schools to build latrines and water supplies was begun, since 25% of the population there go to school. At the suggestion of the Divisional Wells Committee, bleaching powder was sold through a local store. The committee then began to help people chlorinate their wells if they opened them and to sterilise water storage containers routinely. As you may gather boiling water before use is not a practical solution in Kibwezi. But really to consolidate this engineering response what was called for was a very detailed survey of where the coliforms enter the water from the point of collection, through transportation to storage and use. Finding the source of the pollution may stimulate design improvements in the system. It would also identify what hygienic practices would be most effective and worth promoting. But sadly such a survey was not carried out.

Unless the result of drinking contaminated water is both drastic and widespread it is difficult by perception alone to accredit it to a particular source. It is only necessary to consider the situation in a rural school, where children each bring their own water and share it with their friends to realise the significance of Cairncross's concept of "public and domestic domains in the transmission of disease" (8). However, it is possible to help local people to confirm or dispel their belief that certain waters are responsible for diseases of faecal origin by them testing the water with the consumers (6). Such activities where consumers actually culture and see suspected organisms from their own drinking water are of very great value.

Chronic perception. The quality of groundwater may not be constant over time, shallow wells show considerable seasonal variation particularly as regards salinity, whilst the effects of some contaminants such as Fluoride may only become apparent over a long period of time. Whilst the effects of additional contaminants may only become apparent in the long run and when other factors come into play; Trihalomethanes, a byproduct of the superchlorination of water which has an organic content, are suspected carcinogens, however the onset of cancer may be triggered only when additional factors are present. Thus in general the local perception of contaminants which have a chronic effect is at best only a suspicion, unless records of some sort are kept.

The Maasai in Kenya however are well aware of the connection between Fluoride rich groundwater and dental fluorosis. The older generations remember their parents having white teeth, but the present generations see stained teeth in people using borehole water. It is important to note dietary changes since nowadays the Maasai consume more food and drink prepared from the Fluoride rich ground water.

In London, the irradiated rainfall from the Nagasaki and Hiroshima atomic explosions began to appear in tap water after 30 years residence in the supply aquifers to the north.

The chronic effects do not represent an impossible situation, but clearly it is one which has to be monitored and local communities are ideally placed to monitor these trends.

Other facets of perception. I have dealt briefly with what can be perceived about groundwater quality using our senses of sight, taste and smell. Dowsing for water using forked sticks, wires, or a pendulum is a very old art but in Britain it was only in 1952 with the exclusion of dowsing from the witchcraft law that mainstream scientists began to associate with it. There are an increasing number of reliable studies being published on the usefulness of dowsing, whilst in practice water authorities and drilling crews often employ the skill. The accuracy of a dowser is easily determined (9) and dowsing is an effective community-based skill to determine groundwater quality and location. Natural indicators such as soil type and flora can also provide successful clues as to the location, depth and quality of groundwater (10).

We have to consider how we bias our perception. A Maasai community consistently said to us that a well dug on the edge of their clan boundary was too salty to drink. This claim could not be substantiated by repeated water quality testing. It took some long discussions to realise that this group of people would rather do without the well than have a dispute over it with their more powerful clan neighbours. In western Kenya we had finished building a well and were planning the opening ceremony. The chairman's wife agreed to brew some beer for the big day and promptly went and drew water from a local pond because, she said it made much better beer. It may not be a question of removing this bias when we choose to use the results of perception, it is more important to find out why the bias exists.

Thus if we can consider that both scientific and popular perception have a role to play in the improvement of health status and behaviour, even though both can contain bias and personal opinion, it remains for us to consider ways in which we can give local perception the opportunity to enter into the forum of the decision makers. Such opportunities do exist and we should explore them. For example, the local population represents a far more extensive monitoring network than any government could hope to establish. Supporting their interests also generates votes. Cost effectiveness is a dominant yard stick and it would be interesting to compare a community based monitoring system, given the ease of use of global positioning systems, feeding into a GIS type database for ease of data handling, with alternatives for the same situation. In South Africa a project has begun on surface water monitoring which targets schools with excellent educational and investigative resources. Behind the initiative are water authorities, the national parks boards, wildlife societies and universities and multinational corporations finance the venture. The Catchment Action programme has its origins in the Umgeni Water Company in Pietermaritzburg.

As you will gather in my descriptions I have been trying to fit local perception into the scientific view of water quality. Local perception is not particularly effective in identifying specific contaminants, nor can it say much about the quantities involved. Yet working with local perceptions is particularly rewarding in realising the same goals for which we collect scientific data. They are complementary rather than contradictory. The challenge is therefore one of integration.

The Longterm Purpose of Considering Local Perception

In my introduction I implied that the techniques of eliciting and using local perception of groundwater quality are well established and growing, especially in developing countries. But what is of more importance, should we choose to pursue these techniques, is what systems we intend to build them into and for what purpose. I failed to find any reference to local perception of water quality in the past few years journals of the Chartered Institution of Water and Environmental Management from U.K. Yet in U.K, consumer dissatisfaction over water quality and services are reaching critical levels. Consumers perceptions of water quality have a very real purpose they indicate the acceptability of the exploitation of our communal natural environment.

But the term perception strikes discord in the minds of many scientists. We should however consider the history of science before we accept this contemporary view. In 1514 Nicholas Copernicus, a Polish priest who studied in the monastic system where there existed a harmonious line of thought from scientific evidence to belief and religious philosophy, felt it prudent to publish anonymously his view that it was in fact the sun and not the earth which was the centre of our universe. The Protestant Revolution sought to secularize science rather than have it challenge the infallibility of the bible and the scientists were cast out of the monasteries. The scientists sought patronage in the outside world and by the time of the industrial revolution the patrons of science were seeing remarkable returns on their investments. Science regained a predominant position in education. Remarkably the frontiers of our present scientific knowledge seem to be restoring the old monastic harmony of thought.

Stephen Hawking in his best seller "A Brief History of Time" (11) concludes that as science draws closer to a unified theory for the origin of the universe it will spur the popular debate on the "the mind of God", an outcome which Martin Luther would have approved of. Whilst in the realms of environmental science there is talk of a second Copernican revolution in which mankind accepts that we are not the centre of the biosphere (12). Clive Ponting wrote in "A Green History of The World" (13) that environmental thought is the only significant newcomer to the political arena this century and he develops a case that without adequate environmental awareness in the population the best efforts of science are defeated, thus predicting that the success of governments and our existence hinges on popular environmental knowledge. In our society information is ever closer to our fingertips spearheaded by the rapid growth of the Internet and other information technologies.

This crude historical perspective may help us shed light on the purpose of pursuing local perceptions. We try to reconcile our beliefs with scientific evidence, we aim to satisfy the consumers, to "maximise the benefits" and change behaviours. It is the nature of this reconciliation which is important. We can choose to continue to plan for science to contradict popular belief or we can try to plan for coexistence. Within the environmentalists rhetoric there is emerging an increasingly more eloquent expression of the view that harmony of thought promotes harmony in the environment and is central to our own well being. If we choose not to act upon popular environmental thought we begin to pose a threat to our society.

The Longterm Usefulness of Local Perception

Over the past ten years "technology with a human face" became the password to the coffers of the donor community. The vogue to interact with the people, to be in "the field" and to do something "grass roots" has had a fantastic effect. People actually got together and talked. The big shots flew in and the villagers roasted a goat, the global village became a reality. We cannot ignore the tremendous benefits in sharing these experiences. The "development community" has broken down a number of its own elitist barriers and villagers have talked with project planners and "experts" on a sounder footing. It has kindled interest and enthusiasm at all levels.

But how can we continue to make the relationship useful? Do we append the consumers views on the potability of water to the chemists report? Do we go one step further and routinely incorporate this information? Shouldn't we in fact then set up a system to respond to local perceptions alongside any public health monitoring mechanisms we have? Or should we open a complaints department to reply promptly and courteously to the public, showing we are concerned and aim to maintain good consumer relationships? But there has to be caution lest we treat water as a marketable commodity and not as an environmental resource for which we are all responsible.

We clearly have to build on existing initiatives to integrate them into more useful systems. As a starting point for discussion I will propose three possible scenarios.

We could accept that the information gained is for the sake of the community collecting it. Through their own efforts and with some outside help, the people increase their understanding of their environment and improving their health status. Some of this knowledge will become "traditional" and not be lost, and we maintain healthy contact between the outside helpers and the local people. It would increase popular environmental awareness which is of increasing political relevance and the health status of the people would improve. However the information would not necessary feed into any other system.

Secondly we can consider that a more aware and better educated community should be more motivated to replicate development initiatives, they will turn projects into programmes and increase coverage, they would become empowered to realise the goal of safe water for all. But in practise we have seen that this take off point is very elusive without mobilising the essential economic resources. We now talk of trust funds for local water users groups and family groups replacing community groups. We have seen the emergence of a potentially contradictory "demand driven approach" within the development sector. In practise this approach will demand of the development agency that it develops marketing strategy tactics. In effect what was called community mobilisation may become consumer awareness promotion. Undoubtedly what the consumers think is very important in this equation, but the response has to be loaded in favour of the supplier rather than the consumer for the supplier to succeed. Whilst on a more practical note, the NGOs have been successful in the past in their community activities precisely because they are viewed as being there to help rather than to establish a market for a product in which they are a key controller; the NGO's could be trusted. Quite how the NGO's will make this turn around in their managements from givers to dealers remains to be seen. Even so this overall plan for development does not represent a coalition of ideas; it consists of governments, development agencies and the consumers as clearly divided groups. It divides responsibility for what is increasingly seen as a communal commodity and we are apparently also dividing the agents of change into ever smaller units. In the past we have tended to think that in responding to market forces we exclude public interests.

Thirdly we could choose to envisage a perhaps more distant conceptual and unified approach. Under the terms of many finance arrangements for aid projects the sidelining of ministries in favour of NGOs will in the long run only create un workable friction. The very nature of water as a natural resource demands central control; abstraction licensing and effluent standards being clear examples. The NGOs are increasingly receiving the lions share of funding and trying to work with a ministry which they know is essential but under funded.

There is presently an initiative under way in the Great Lakes region of North America, which sheds light upon this difficult situation. The Great Lakes span the Canadian-U.S. border. The pollution issue; the U.S. automobile industry on the lake shores, is fraught with rhetoric and political posturing. The area concerned is enormous and contains around 30% of the worlds fresh water reserves, there are major cities and large tracts of sparsely populated land, it is clearly emphasised that under these conditions a central government systems cannot hope to have the resources to police the water quality entirely. The responsibility for environmental protection is self evidently our responsibility, we cannot assume someone else will look after it. A view increasingly promoted in Canadian environmental circles. Communities in the region; urban, native American and Innuit have been given access to information and encouragement allowing them to organise themselves and prepare evidence on pollution issues (14).

Via community groups, environmentalists, and liaisons with government offices and universities; the local people have at least a system through which they can report to government on their environmental concerns. The reporting is not just a source of extra data, it is part and parcel of the environmental monitoring policy. Of course the bottom line for such an approach is the availability of an abundance of political will.

Perhaps it is something along the lines of this triumvirate of local people, technical agency and government with a mutually agreed responsibility for tackling water quality issues which could determine useful long term collection of local perception information. The people live where the information has to be collected, the government safeguards the well being of the people and the people choose their government. Preservation of the environment is on common ground.

Conclusion

Local perception of groundwater contamination is as useful as we choose to make it. There are numerous examples of well thought out investigations of local perception of groundwater contamination improving our understanding of a situation and implementor - beneficiary relationships. It remains for us to devise purposeful longterm systems to build upon this starting point. These systems could be built around an increasing awareness that thorough environmental contaminant monitoring extends beyond the resources of formal government networks. That local people represent a much better placed network and their environmental awareness constitutes an ever more important political lobby. Local Perception can also be seen to form part of a continuous line of thinking and belief which not only contributes to greater scientific understanding, but is also the very basis of our health related behaviour. Our present systems appear not only to partition knowledge between local perception and scientific data, but also to partition what is a communal responsibility to safeguard a natural resource.

References

1. Minutes of a Meeting of Mombasa Water Committee, 14 Sept 1908. *Kenya National Archives* PC Coast 1/14/103.
2. Tebbutt, T.H.Y. 1983. *Principles of Water Quality Control*, Pergamon 3rd edition.
3. WHO. 1985-86. *Guidelines for Drinking Water Quality. Vol 3. Drinking Water Quality Control in Small Community Supplies*, Geneva.
4. Woodhouse, M. 1995. *Salinity and Shallow Wells a view from the field. in Groundwater Quality*, Chapman and Hall.
5. Lehumsluoto, P. 1987. *Provision of safe potable water in rural areas.* African Water Technology Conference Nairobi 1987 and University of Tampere Finland.
6. Woodhouse, M. 1990. Consumer involvement in bacteriological testing of well water in Kibwezi, Kenya. *Waterlines,* v8 no4.
7. Musomi, J.M. 1992. *Water Protection and Sanitation in Relation to Faecal Coliform Contamination in Ngwata Location Kibwezi Division.* MPH Thesis University of Nairobi.
8. Cairncross, S. 1995. *The Public and Domestic Domains in the transmission of Disease. People and Pathogens* Conference Eldoret Kenya.
9. Woodhouse, M. 1992. Locating Wells Using Dowsing and Earth Conductivity in Kenya. *J. British Society of Dowsers* Volume 35 No 235
10. Woodhouse, M. 1991 Natural indicators of ground water in Kibwezi Division, Kenya. *J East African Natural History Society and Museums.* Vol 81 No 197.
11. Hawking, S. 1988. *A Brief History of Time.* Bantam.
12. Taylor, D.M. 1994. *Off Course, restoring the balance between Canadian society and the environment.* IDRC Ottawa.
13. Ponting, C. 1991. *A Green History of The World.* Penguin.
14. Great Lakes Health Effects Division, Statistics Canada.1994. Draft handbooks on investigating human exposure to contaminants in the environment.

THEME V : ARE PERIURBAN SLUMS TO STAY WITH US?

Chairman: Mr. K. Kamar

Paper 12. HISTORICAL CONSIDERATIONS WITH REGARD TO THE IMPACT OF KNOWLEDGE AND ACTION ON WATER QUALITY RELATED BEHAVIOUR

Christopher Hamlin
University of Notre Dame, Indiana 46556, USA

As an historian of science I have been much concerned with the generation of knowledge, about water. But as an historian of public actions to secure better water, I have discovered that more important than knowledge is meaning. Let me illustrate with an anecdote. I grew up in a small farming village of about 500 people. We had a public waterworks of some sort, with water taken from wells about a mile up the valley. It was a water of variable quality; almost every spring during the snow melt my parents would announce that they had been told it was time to boil the water, and we would drink tasteless flat water for a month or so. I became more familiar with the water quality when I was a teenager in secondary school, and found that mother of a friend - she was a school nurse, and the only medical professional in the town - was doing the monthly coliform counts. I remember on one occasion the count was unusually high. But far from being concerned, we were secretly delighted.

We ourselves were used to the local microbial ecology, but we had heard that frequently when rich folks from the cities - New York particularly came to visit our rural paradise to stay in their luxurious holiday properties - that they suffered a few days from intestinal discounts while getting acclimated to the water. Certainly we didn't wish them serious harm, but we thought it served them right to suffer a bit. After all, they were sophisticated and wealthy, with all kinds of advantages we didn't have - a little diarrhoea evened things out. We felt victimized on their turf; it was only fair they should have a little difficulty on ours. Just as their humour made fun of us as country fools, ours made fun of them as city slickers - getting lost, stuck, being petrified of dairy cows, etc. Had someone appealed to these high coliform counts to propose that the water be chlorinated or filtered I suppose we would not have objected very much (except perhaps to the expense), though we would have seen it as highly unnecessary. I do not think this attitude is idiosyncratic.

In the U.S. there used to be a phrase "Montezuma's revenge", referring to the diarrhoea that many US citizens experienced when they went to Mexico. Montezuma was the chief of the Aztec nation destroyed by the Spaniards in their early 16th century conquest of the Americas. I do not know that many Mexicans identified with Montezuma or felt a slight sense of getting even when tourists suffered, but the term "revenge" suggests that some balancing of right and wrong is going on.

It need hardly be said that the secret delight at the prospect of another's suffering was not what Robert Koch was intending in developing the procedure of counting coliforms as a test of water quality. He assumed that his contribution belonged to a rational system of scientific problem solving.

The relation of knowledge to action

At issue here is the relation of knowledge to action, which is something we really know little about - at least in the broad sweep of history. We see advances in knowledge; we see efforts to improve water supplies, and we assume that the former is sufficient cause of the latter. Accordingly in the histories of water and health, advances like Koch's loom large, for they are assumed to be the determinants of progress. But the more closely we look we find groups of people who did not wait for Koch (or John Snow) to improve their water, or people who, like my boyhood friends, know about the effects of water on health, but don't choose to act in the way Koch would expect them to. The reasons why water safety is achieved in some places and not in others (and the way that scientific knowledge fits into those achievements) turn out to be very complicated, and often dependent on personalities, local politics, the ability to forge alliances, and so forth.

While many here may have experienced some of that complexity, my sense is that the kind of perspective represented by Koch still exemplifies what "public health" is. That is, there is a professional identity that is manifested in a reading of history. In any profession, we appeal to history when we think about how we define our domains of expertise, what courses of training we insist on, how we negotiate our job descriptions, set goals and work toward them, and finally what expectations we have about the good we can do. Our very ideas of a professional trajectory - of where our profession has come from, where it is going, and why it must promote certain standards and modes of activity - are neither more nor less than an historical interpretation. The people who give us that interpretation - our teachers, editors, textbook writers; the presidents of societies, or the writers of mission statements or annual reports - are acting as historians, whether intentionally or not.

What sort of histories will best serve the profession of public health, and particularly those concerned with water quality? Public health is a hybrid profession. It is connected with medicine, but also with sanitary engineering, with chemistry, biology, and police and public administration. In part it is scientific, in being concerned with the increase of knowledge, but more immediately it is goal-oriented, being concerned with the discovery and prevention of disease. In many places, it is a profession continually under strain. For many reasons - because public health personnel are often seen as interfering with customary life and making things more difficult, because they are being most successful when their services seem least necessary, governments, at all levels, can often find more urgent priorities for their money and attention. The several articles in a recent book by Fee and Acheson on the History of Education in Public Health reveal a beleaguered profession continually having to redefine itself to meet the changing whims of governments.

The Koch exemplar focuses attention on public health as science. In the past, the link to science, and a presumed link of science to humanitarianism - has been one of the chief means the public health profession has used in its struggle for recognition. It has called upon what one might call "superstition-enlightenment" history. This history functioned in several ways. The appeal to science suggested that nature itself prescribed the principles and skills of public health, and united the profession - it justified rules stipulating the training and responsibilities of public health officers. The strong association with humanitarianism helped the new profession link itself with various broader social movements, like progressivism in government, or social reform, or, earlier, the evangelical revival. It attracted idealistic young people to the profession; it reassured them, even in the dismal days of running county hospitals, that they played a unique role in the history of civilization, as the bringers of health, that foundation of so many (though not all) concepts of human well-being. Most importantly the link with science was the warrant for the new profession's claim of authority, a claim all professions must be able to make. With the aid of our microscopes and our culture plates, with our mortality statistics and case-tracings, our eyes alone can see the invisible horde that causes disease. Believe us, do what we say, and we will protect you. This is very much the overwhelming message of the early days of the germ theory. It was a very important message, an essential one for public health to gain a foothold in public service.

You may think from the way I have developed this discussion that I am now going to criticize this history as inconsistent with the historical record or something like that. I am not. These functions continue to need doing; it is essential to have a history to help do them. What I do want to suggest is that this history needs supplementing. What I have described are the ways history aids in professionalization, in uniting a profession, getting it respected and prominently placed in public life. But that history doesn't necessarily help the profession accomplish its goals in changing water supplies. For that, we need a different kind of history. In the case of water quality, this new history must answer two kinds of questions: 1) When water supplies have been improved what factors have motivated people to improve them? I'll use a broad concept of improvement here - including increased quantity, or easier access, or improvements in taste, appearance, or in properties that make it more useful for particular industrial processes, as well as in the more specific area of hygienically purer water. And 2) What has been the role of "knowledge-authorities" (scientists, public health professionals) in this process?

Before I begin to answer this question a few more points about "superstition-enlightenment history". It rests on a range of assumptions about the relation of knowledge to action which can be labelled the rationality presumption. It assumes that when problems arise, problem-solving efforts slowly gravitate towards them, like macrophages toward invading microbes. Ultimately, they produce the knowledge of the right solution, and once the right solution is found it diffuses to all places where that problem exists, and stimulates the correct responses, and so the problem is then solved everywhere. The metaphor of enlightenment itself envisions knowledge spreading out like light and illuminating wherever it hits. That metaphor makes a big distinction between the knowledgable few and the ignorant many, who are looked upon as miserable, helpless, and passive until they have been enlightened, and who will respond in exactly the same way to enlightenment. The rationality presumption assumes further that solved problems stay solved. Rather than having to re-fight old battles, the problem-solving institutions, strengthened by their past successes go on to tackle the next generation of even more difficult problems.

For a whole range of reasons things often don't happen that way. Take, for example, the classic water quality case, that of cholera. This is the exemplary case for "rationality" history. After the first pandemic the nature and causes of cholera came to rest on the research agendas of European medical scientists. By carefully studying what those who got cholera and alternatively those who didn't get it, had in common, one of these scientists, John Snow, was in 1854, able to demonstrate that cholera was a faecal-oral disease transmitted by water. Following Snow, efforts were made throughout the world to safeguard water supplies from contamination not just by that cholera vibrio but by all manner of pathogenic agents, biological and chemical. If knowledge were enough there should be no cholera and no other water borne ailments. But there are. Some of the reasons, of course, have to do with the wiliness of microbes which can change and hide, and the appearance of new kinds of agents that cannot be dealt with by existing technologies. Yet often societies seem also unwilling to keep up institutionally, to implement the technologies that exist. In many places in the United States, the quality of public water is perceived to be getting worse. As the very institutions, the public water services, that exist to protect health are perceived as providing a tainted product (and of lying about its quality) their credibility is undermined. The problem here is not ignorance - the experts know what the problems are. We can say the problem is a callous disregard for health on the part of the public authorities which must authorize improvements, but that doesn't help much.

Why have people acted ?

On then to the first question, why have people acted to improve water supplies. Standard histories of public health usually see concern for water as contemporaneous with, and a result of, the revolution in sanitation, with the rise of knowledge of faecal-oral disease transmission. Prior to that time societies (with the exception of ancient Rome) are usually presented as little concerned with the provision of water, and virtually unconcerned with water quality. This is not true. Historical records indicate a very widespread, one is tempted to say universal, interest in good water that long predates the famous nineteenth century bacteriological and epidemiological discoveries. It is important to inquire into the bases of these concerns because they may suggest strategies to improve water that may sometimes be more effective than hygienic education alone.

I am going to consider briefly six reasons for concern about water: health, aesthetics, local pride, prestige, profit, and social control. To appreciate some of these at least we need to recognize that water has not always meant what we mean by it: a substance that is necessary to ingest in some form, highly useful also for cooking and for washing person, clothes, and other household items, that also can serve as a vehicle for transmission of pathogenic and toxic agents. It is treacherous to generalize, but a common theme in classical writings in mediterranean Europe, the near east, and China, and in collections of European folklore, is to speak of "waters" rather than water. Not only is water seen to vary enormously in terms of origin (rainfall, snow-melt, dew, pond, spring, and river water were seen to be significantly different), waters also seemed to be unique features of place. The best analogy is to fine wines now, as bearing the unique characteristics of their place of origin. While the modern taxonomy of water quality includes only two classes (pure and impure), subtle and complicated characterizations appear nearly universal in the past, especially with regard to water from springs. But these were not just more complicated inventories of qualities. We are wrong also to assume that people invariably recognized waters as a class of material substances, varying in containing greater or lesser quantities of dissolved or suspended contaminants. That concept, of one chemical substance included or dissolved in another would have been incompatible with the long dominant Aristotelian natural philosophy. These unique waters are often better understood as quasi-spiritual entities, each having a unique "life" of its own. That perspective arises very clearly among historians of comparative religions and mythologies, like Eliade, who find universal themes in water symbolism from peoples throughout the world.

On then to ***health***. We would be wrong to think that people had not arrived at notions that certain waters were unwise to drink before the discoveries of John Snow. Classical writers - Vitruvius, the Hippocratic texts, Pliny - lack the concept of a clinical disease entities, but are full of reports about the adverse effects of drinking particular waters. Often the advice does not refer to anything recognizable as enteric disease, but to the bizarre effects of waters on fertility, or on one's singing voice, or the colour of a sheep's wool. Nonetheless, Avicenna notes that travellers must take special care in choosing water: "The traveller is more exposed to illness from the diversity of drinking water than he is from the diversity of foods....it is necessary to be particular about correcting the bad qualities of the drinking water, and expend every effort in purifying it" (Gruner 1930: 454). Several forms of purification are spoken of: boiling (Avicenna), or the addition of vinegar or wine, or the soaking in the water some substance such as pearl barley, onions, or wax, which would absorb or counteract injurious matters, or the shaking of a suspect water with sand, or using a wick of fleece to transfer water from a container, or filtration, either naturally through soil, or artificially through wool, bread crumbs, or cloth. (Gruner 1930: 222, 454-5; Baker 1948: 1-8; Lorcin 1985: 263).

Concerns about effects on health merge into ***aesthetic concerns***: taste, particularly sweetness, appearance, digestibility or "lightness". Lightness did not usually refer to weight, but to effect on stomach, but some medieval authors did suggest that one could measure it objectively: one could wet equal areas of tissue in different waters and compare their weights to determine which water is the lighter, and hence the better to drink (Lorcin 1985: 262). The classical literature, eastern and western, gives enormous attention to questions of when to drink particular waters (e.g. before, during, after meals), to how much to drink, to what the effects would be. For example, upon waking with a thirst, one should resist the temptation to take water: it would not satisfy the thirst, and would cause indigestion and bad dreams. Nineteenth century sanitarians argued that lay persons' judgments of water quality should be discouraged: not only were they unreliable guides to detecting harmful water, they might actually lead people to favour bad water, as in notorious cases where people went out of their way to the sprightly water full of nitrates from decomposition products.

In fostering the view of lay incompetence in judging water supplies, were they perhaps going too far? In many more cases, however, the judgments made on grounds of taste, appearance, subacute physiological effects were probably wise judgments, and in some cases, particularly in instances of episodic or unexpected contaminations of public water supplies, it has been taste that has first brought the problem to notice. It is one thing to say that the judgments of lay people should be cautious and educated, quite another to announce that lay people are utterly incompetent and that their aesthetic views of water quality don't matter. Made to believe their perceptions are valueless, they are likely to become uninterested in bringing those perceptions to consciousness, or in being concerned about water quality. On the other hand a population acutely aware that water does have varying properties that they can judge is likely to generate an interest in enhancing those qualities. I see a sharp contrast, for example, between the ways in which classical authors talk about the subtle effects of exotic waters (like the drips from stalactites) and the attitude of most people I know, which is that at least in the civilized world, all water is the same - in taste and quality: all is equally uninteresting. Having been told that they no longer have the right to appreciate water quality, people must become even more dependent on an administration of experts, with which, for a variety of reasons, they may not want to cooperate, and which may in fact not warrant their trust.

Closely bound up with aesthetic judgments and notions of the uniqueness of particular waters are notions of ***local pride*** in one's water. Ideas of the individuality of sources of water are most clearly evident with regard to springs and streams. In many European folk traditions the individuality of streams and springs was attributed to spirits or resident divinities. Best known are the thirty thousand nymphs associated by Hesiod with springs in Greece (their brothers were the rivers) (Hesiod V, 337-82; Moser 1990; Tolle-Kastenbein 1990). The efforts of the medieval Catholic church to halt worship of such divinities usually failed; instead they led to the association of wells and springs as the sites of miracles associated with particular saints (Hope 1893; art. "Brunnen" 1927; Vaillat 1932; Guitard 1951; Bord and Bord 1986; Guillerme 1988).

One can say in retrospect that medicinal properties of spring waters may account for their association with divinities. Perhaps that is so, though it would appear that many more springs and rivers had divinities associated with them than we would characterize as having some unique "mineral" character. What is important is that the association of a specific divinity with the unique life-sustaining properties of the water of each locality spring tended to give that water and that place, a cosmic significance, a special goodness which local residents must feel, take an interest in, and sustain. Wild, for example, has argued of the Nile-worshipping cult of Isis and Sarapis (important during the early Roman empire) that Nile water was simply understood as the most perfect water; it was associated with fecundity, for humans as well as for crops, and was known as a fattening water (Wild 1981). The state of that water would then be a measure of the quality of the place in a cosmic sense - did it still enjoy divine grace. There would be as much reason to take an interest in that water, to celebrate it, to try to protect it, as there is for a middle class modern American to determine his cosmic well being in the state of his portfolio of investments.

Over the centuries this sense of specialness of place became secularized. Residents discovered that the promoting the superb qualities of local waters - whether for feeding stock or treating invalids - was a good way to make money. Still, even into the nineteenth century, they often talked semi-mystically of the unique "life" of waters. Even when analysts found nothing in those waters to account for their supposed distinct properties, they might still be recognized as indicating a special grace that lay beyond the dissecting powers of science. In Britain in any case one still sees holdovers of spring worship in well-dressing ceremonies. Similarly, the revival of the bottled water industry has kept this kind of sensibility alive, just as the promotion of spas did for many years. In this regard, Goubert's observation that in the nineteenth century water became simply another commodity, is important. Commodities are items whose origins and sacred significance doesn't matter -they are completely interchangeable. We trust authorities to insure their basic safety, and usually take no further interest. Even with the best intentions, the authorities don't always merit that trust. My point here is simply that to regard a water source as something unique, special, providential, offers a much stronger motive to take an interest in its protection than to suggest that the water is just ok.

Prestige. But celebration of place or of water, doesn't just happen automatically. One of the things which is very clear in the history of the provision of good water is the attempt to use water supply in an effort to create a public memory or a sense of community. The clearest pre sanitary revolution examples of this are memorial fountains. Most of the interest in these has been by architectural historians. Yet such fountains did serve as the water supplies, particularly in central and southern european towns. Sometimes, as in 16th century Berne, the patron is the town itself, and the fountain is topped by the arms of town. Often the patrons are wealthy aristocrats who want to be remembered as philanthropists. It is true that these fountains did not serve to get water into dwellings, and that for the same investment required to build one grand fountain one could have brought water much closer to homes. It is well to point out that public fountains can be ideal sites for disease transmission. Yet one must also note that fountains were often supplied with water taken from well outside the town and that many of them clearly have bucket filling apparatuses which allow you to get water before it mixes with whatever has got into the trough. During the late nineteenth and the early 20th centuries, when towns took over the responsibility for supplying water, and when they switched from building magnificent fountains to reservoirs, filtering and pumping stations, water towers, and systems of mains, this idea of performing a monumental act that would live for ever in public history and knit together the community past and present, did not disappear. While much of their effort was unfortunately hidden below ground, the architecture of waterworks buildings was invariably monumental as well as functional. Something of the communal significance of these waterworks technologies is evident in accounts of the ceremonies at the openings of sewage treatment plants or other waterworks: parades, bonfires, banquets, fireworks, endless speeches. Everyone could take pride in their gift to posterity; the builders and managers were local heros.

In a broader or narrower sense, many of those who led efforts to get better water were seeking to ***profit*** from that activity. The eternal fame achieved by the builder of a fountain to which people will go for centuries hence to cool off, wash, draw household water, is one kind of profit. In the case of municipal water supplies, one often finds that those who lead the often very costly efforts to build public water works stand to gain in more material ways. Silverthorne, the most knowledgable chronicler of the waterworks efforts of 19th century British towns, argued that industrialists, needing large quantities of good quality water for the industrial plants, were usually the prime movers. In many cases their leadership allowed them access to water sources which individually they could never have afforded; in addition they were, as large users, often also able to negotiate lower water rates, which effectively amounted to public subsidies of their industries. But a whole series of others stood to gain too, many of them simply from the building of the works, regardless of how well it performed: landowners, engineers and architects and plumbers, lawyers and bankers, and investors, and labourers. In current American political jargon "water project" means some scheme to use public money to increase employment in the district of whatever congressman is sponsoring it. In order to secure public assent to build the project it is necessary to talk about benefits, in terms of household water quality, or acres to be irrigated, etc, but in practice these benefits are often negligible, and arguably, to the creators of the project, they are secondary to the creation of jobs.

The final example of a motivation for improved water is ***social control*** - what might more fairly be called the use of water to change social relations. For the noted sanitarian Edwin Chadwick this, and not disease, was a primary concern. As well as believing that dirtiness of house and of person was the main source of moral degradation and ultimately revolution, Chadwick was particularly concerned with what he regarded as dangerous forms of socialization that took place in water queues. It was not that water queues represented a tedious and exhausting mode of water supply, but what people, particularly women and children did to relieve the tedium. In water queues, people squabbled, children learned bad language, assignations were made. The same concerns are reflected in the sanitarians' ambivalence about public wash houses. They appreciated that even with piped in water the tenement was not a salubrious place for laundering, but saw the gossip of the public wash house was an even worse alternative. Chadwick also argued that the time saved by not having to fetch water could then be converted into productive time, allowing factory hands to stay longer in the factories, earning more wages for themselves and greater profits for employers. One need not share Chadwick's social goals, one need not think that socialization among women is politically dangerous or think that the proper thing to do with saved domestic labour is to lengthen the hours of wage labour, to recognize the advantage of multiplying the number of beneficial ends that will be served by the same action. Modern commentators have noted that social factors, like the lack of concern among male community leaders in technologies whose benefits would be most striking for women, do impede the securing of better water supplies, but in the same fashion one can see the possibility, in situations where there are powerful forces to improve women's opportunities and education, to link attempts to improve water supply to that movement. What Chadwick shows us is that alliances among reform movements with quite different goals may mutually reinforce one another, achieving results which none of them might achieve urged on its own. I would argue that Chadwick's ability to appeal to matters well beyond disease or even dirt - to matters of social stability and economic productivity - were crucial in the ultimate success of his campaign, and indeed that to many in power the latter concerns seemed a good deal more immediate than the former.

Sir Edward Frankland

Can water professionals - the knowledge authorities I spoke of earlier - engage fruitfully with these kinds of incentives to improvement of water? Doubtless many do. One who was extremely adept at turning knowledge into action was Sir Edward Frankland (1825-1900), an English chemist, and the world's leading authority on water quality from the mid-1860s to the late 1880s, roughly the period after John Snow demonstrated the water borne transmission of cholera, and before there was substantial bacteriological experience with water borne pathogens. During that period Frankland was England's premier chemist, as professor of chemistry at the Royal College of Science in South Kensington. Early in his career, around 1850, he had made important contributions to the theory of valency, and done important original work on organometallic compounds, and this early work was the source of his fame among chemists, but for the main part of his career he was not an academic but a public chemist, interested in many things but specializing in water and sewage.

Frankland got involved in water accidentally. In 1865 he became the South Kensington professor when his predecessor, Hofmann, was suddenly offered the prestigious chemistry chair in Berlin. With Hofmann's chair he also inherited the minor post of water analyst, which required a monthly analysis of the water supplied by each of the eight companies that served metropolitan London. Early in 1868 when the government sought to reconstitute a royal commission investigating British river pollution (its former members had been unable to agree on the relative merits of water closets and various forms of pail closet), Frankland was an obvious choice to be the chemist on the commission. The commission spent much of the next six years investigating problems of water supply and sewage treatment as well as river pollution. Initially there were only three commissioners. One soon died, and the other was not particularly interested or expert in the matters at hand, so for practical purposes Edward Frankland was the Royal Commission on the Pollution of Rivers. That appointment made Frankland the director of the largest water laboratory in the world (probably the only one exclusively dedicated to waters) and effective director of an investigation that received roughly the same amount of funding as the British government awarded to the Royal Society to subsidize scientific research in general. Frankland used the lab well, to refine processes for the chemical analysis of waters, to carry out definitive studies on the self purification of rivers, to develop methods for recycling manufacturing effluents, and to develop effective intermittent filtration sewage treatment methods that vastly increased the amount of sewage that could be treated on a given amount of land.

But Frankland's most important achievement was not anything he discovered, but the way he reacted to what he could not discover, and in particular, the way he conducted the public relations of water supply. A key turning point was in the summer of 1866, during Frankland's first year as water analyst. London was hit by cholera which was clearly being transmitted by water. To his surprise, Frankland found that the cholera causing water was analytically indistinguishable. Indeed, according to conventional ways of measuring water quality, which focused generally on the concentration of organic matter in the water, diluted cholera evacuations could be shown to produce a better analytical profile than ordinary good water.

What is important is the way Frankland reacted to this discovery. Keep in mind that he was working during a period where there was significant, though not overwhelming concern about water quality, based on epidemiological demonstration, within the public health community, and a good deal of speculation about cholera "germs" - the term was simply a placeholder for the whatever-it-was that got in the water and caused disease. Culture methods existed but were rudimentary, and there seemed no good reason to think that their use would ultimately lead to discovery of the cause. Indeed, there was no obvious research program to distinguish cholera-producing water. It seemed impossible to discriminate different types of contaminants; simply to measure the organic load was at the limits of analytical chemistry.

Given this dilemma, Frankland might have done several things. Had he been confident that chemical analysis was sufficient to the task of determining water quality, it would have been intellectually honest to have rejected the epidemiological inference. Since chemistry showed the water was not bad, the cholera must have some other cause. Or one could assume that chemistry was good, but the contaminating events had been brief, and eluded the sampling. Alternatively, one could simply have confessed that chemistry was powerless to disclose water borne disease. Hofmann had taken such a view. He hoped it would lead to the adoption of new sources of water that could be protected from contamination, since no one could say whether the existing sources were contaminated. But his knowledge was not granted that meaning: instead, the public took Hofmann's failure to find anything clearly wrong with the water as proof that it was safe. If Hofmann, who, after all, was the expert, could find nothing to object to, why should ordinary citizens second guess him.

Frankland shared Hofmann's belief in the inadequacy of chemistry, but he was much better at translating that knowledge into public concern. He was able simultaneously to claim expertise and ignorance. His claim of expertise was that he knew more about water quality than anyone. The public should pay no attention to backroom analytical chemists who might tell them their water was fine, because he, with a far more sophisticated laboratory, and working orders of magnitude beyond their capacities, knew that they were incompetent. Yet while he knew far more than these others he knew nothing about whether water was safe.

One might assume that the response to this kind of pronouncement would be utter despair, municipal officials throwing up their arms that the greatest authority, the only one whose position indicated impartiality, could not answer their key question. But Frankland made sure this did not happen, and he did so by adroitly reinforcing many of the social forces I considered earlier, which would tend to encourage people to adopt better water. For while he was not prepared to say that a given water sample could not cause cholera, he did have very clear ideas about which kinds of water sources were likeliest to be safest, and which kinds were likeliest to be dangerous. These ideas were pretty commonsensical. Water from very deep wells, or from areas above human or animal habitation was likely to be pretty safe, water contaminated with the wastes of humans or animals was likely to be dangerous. What Frankland was saying was that ordinary people familiar with the geographic circumstances of different waters, usually knew a lot more about their probable safety than he did from the sample bottles they sent to his lab. But to have urged people to use their common sense would have left them prey to other experts who would dazzle them with numbers and arguments that could circumvent these obvious truths.

So Frankland kept up the illusion that his meticulous analyses meant something, that they drove his judgments. All the while he presented his results in ways that reinforced lay concern, sometimes by carefully managing public anxiety. In 1868, for example, he incorporated into his tables a new category for stating water quality, previous sewage contamination. Previous sewage contamination was nothing more than a manipulation of measurements that appeared elsewhere in the returns. One took the total nitrogen breakdown products in the water (nitrates, nitrites, and ammonia), calculated the average amount of organic sewage matter that would have yielded these products, and then reintroduced the average dilution factor of London sewage to yield a figure, which could be expressed in terms of pounds (usually several thousand pounds of previous sewage) per hundred thousand pounds of water (Frankland thought using the customary measurements of commerce was much more intelligible to people than the scientist's parts per million). Critics protested that this was highly misleading: according to the pathological theories of the day, the fact that the nitrogenous materials (which might not all have come from animal or human contamination) had completed their decomposition was proof of their safety. Quite so, agreed Frankland; the figure was just an approximation, after all, and perhaps they might not always complete their decomposition. The effect, however, achieved by reporting the levels of previous sewage contamination in the newspapers each month, was to plant in the public imagination the truth that one was drinking sewage.

Frankland similarly used to advantage the idea of the germ. True, we knew nothing about what germs really were, but it was plausible to think that they were extremely virulent and extremely hardy. It might take only one to start a case of cholera; that one germ, reproducing in the intestinal tract of the first victim could spawn an epidemic.

Appealing to concepts of these, making the most of his experience as royal commissioner, Frankland made a career of intervening in the attempts of towns to improve water supplies. British water policy during the period typically involved towns decided that they needed better water, developing plans to exploit particular sources, and then appealing to Parliament to pass a private bill that would give them eminent domain privileges to build the necessary works. Parliament's decisions had to be represented in terms of weighing public goods and private inconveniences. Quite often the towns faced opposition, and the opposition's argument was usually that existing supplies were good enough. Sometimes, when this argument came from existing water companies, it reflected an attempt to raise the monetary settlement they would receive when they were bought out (for existing plant and future dividends); sometimes the argument came from landowners who really hoped the schemes would go through, but thought that with such arguments they could intimidate the promoters into buying their land at a far higher price than Parliament would otherwise have awarded. Frankland almost always appeared for the promoters. He became one of the most expert of expert witnesses, able always to reinforce the idea that better water could be had, always to undermine confidence in the status quo.

We might expect the arrival of bacteriology to resolve this situation. Finally, people would be able to make informed water policy. Ironically, when culture techniques and coliform counts became available in the mid 1880s, they were invariably used to counter Frankland's arguments. It made little difference whether such analyses disclosed many or few bacteria. Whatever the results, the very doing of science seemed to signify security, that someone was in control. A finite danger could almost always be minimized. Frankland's indefinite and undefined dangers could not be so readily dismissed.

What Frankland offered was less knowedge than authority. He recognized that he was working in a situation where an informed and attentive public, familiar with the sources of its water and alert to changes in quality, was a much better trustee of water quality than were government agencies or groups of professionals, be they doctors, chemists, or bacteriologists. He was able to exploit the fact that expert and layperson really objected to the same things in water: they objected to water that was demonstrably "foul" or that had been "befouled" - the term catches the right mix of technical and visceral. His involvement was essential; in part because he provided the arguments necessary to demolish the overly sanguine assumptions of those tempted to rely too much on inadequate techniques, in part because he could reassure concerned public by translating their concern into the authoritative language of science. Without it, the citizens groups who had managed to convince themselves of the need for better water, could not have stood up to the experts, who saw their role as superseding lay sensibilities, by pronouncing, for example that while the water might have been fouled, they could certify it now to be pure. On the issue of water quality, Frankland recognized that unless one could ensure that a water source was always free of contamination (which was something these analysts could not guarantee), the fact that it might be safely drunk at any given time signified nothing. In most cases, water borne disease was after all a rare event. To conclude that a water was safe on the basis even of many measurements was as absurd as saying that a house was burglar proof because a house was not being burgled at any given time.

In terms of knowledge, Frankland was a leader. He commanded a knowledge of the literature, an arsenal of data, and a laboratory capability that was unparalleled. Yet this knowledge had a relatively restricted role in the generation of action. It allowed Frankland never to be embarrassed by the superior knowledge of a rival witness, and thereby to maintain his authority. But some fairly simple principles of water safety, rather than vast assemblages of facts, actually guided his decisions about what to recommend. To base decisions on fact trusted too much to the inherent incompleteness of science.

He was effective in action to get better water, not because of all the many things he knew, but because of his appreciation of the political psychology, or public relations of water projects. He knew what to say to strengthen the will of those trying to improve water. He knew how to generate an animating concern for better water, and how to sustain that concern without it lapsing into despair at the impossibility of absolute safety.

Frankland has not loomed large in the history of public health because that history focuses very much on the scientific foundations of public health. Unlike Snow or Koch, he made now distinct advance in new knowledge. On the contrary he was developer of a technique of water analysis that used alone gives one virtually no useful information about whether drinking a water will cause a disease. From the perspective of this other kind of history, a history of action, of securing better water supplies, he is the most important figure of the period because he knew how to convince democratic publics that they needed better water.

References

Baker, M. N. 1948. *The Quest for Pure Water: The History of Water Purification from the earliest Records to the Twentieth Century.* New York: American Water Works Association.

Bord, J., and C. Bord. 1986. *Sacred Waters: Holy Wells and Water Lore in Britain and Ireland.* London: Paladin.

Brunnen 1927. "Brunnen". *Handwörterbuch des Deuteschen Aberglauben.* editors E. Hofmann-Krayer, and H. Bachtold-Stäubli. Berlin: Walter de Gruyter.

Chadwick, E. 1965. *Report on the Sanitary Condition of the Labouring Population of Great Britain* edited with an introduction by M.W. Flinn. Edinburgh: Edinburgh University Press.

Eliade, M. 1958. *Patterns in Comparative Religion.* translator Rosemary Sheed. New York: New American Library.

Gruner, O. C. 1930. *A Treatise on the Canon of Medicine of Avicenna incorporating a translation of the First Book.* London: Luzac.

Guillerme, A. E. 1988. *The Age of Water: the Urban Environment in the North of France, A.D. 300-1800.* College Station: Texas A&M University Press.

Guitard, E. H. 1951. *Le Prestigeux Passé des Eaux Minerales.* Paris: Société d'Histoire de la Pharmacie.

Hamlin, C. 1990. *A Science of Impurity: Water Analysis in Nineteenth Century Britain.* Berkeley: University of California Press.

Hesiod. 1953. *Hesiod's Theogeny.* trans N. O. Brown. New York: Liberal Arts Press.

Hope, R. C. 1893. *The Legendary Lore of the Holy Wells of England: including Rivers, Lakes, Fountains, and Springs.* London: Elliott Stock.

Lorcin, M. T. 1985. Humeurs, Bains, et Tisanes: L'Eau dans le Medecine Medievale. *L'Eau au Moyen Age.* 259-73. Publications du Cuer Ma, Universite de Provence. Marseille: Editions Jean Laffitte.

Moser, H. 1990. Brunnen, Bründl und Quellen: Eine Volkskundliche Betrachtung. *Wasser: Ein Versuch.* eds. G Dienes and F. Leitgeb, 234-6. Graz: Leykam.

Tölle-Kastenbein, R. 1990. *Antike Wasserkultur.* Beck's Archaeologische Bibliothek, Munich: Verlag C.H. Beck.

Vaillat, C. 1932. *Le Culte des Sources dan la Gaule Antique.* Paris: Librairie Ernest Leroux.

Vogel, H. 1988. Brunnen und Pumpereien in der Stadt Bremen. *Wasser. Zur Geschichte der Trinkwasser-versorgung in Bremen.* Ausstellungskatalog Bremer Landesmuseum für Kunst- und Kulturgeschichte. 50-67. Bremen

Wild, R. A. 1981. *Water in the Cultic Worship of Isis and Sarapis.* Leiden: E.J. Brill.

Paper 13. WATER, SANITATION AND THE HOUSEHOLD ENVIRONMENT IN ACCRA, JAKARTA AND SAO PAULO

Gordon McGranahan, Stockholm Environment Institute, Box 2142, S-103 14 Stockholm, Sweden

This paper presents selected findings from a study of the environmental problems facing households in Accra, Jakarta and São Paulo, coordinated by the Stockholm Environment Institute.[1] The emphasis will be on water and sanitation, with some mention of other related household and neighbourhood level environmental problems, such as solid waste, insects, poor housing, and even indoor air pollution.

These many environmental problems are closely interconnected especially in low-income periurban areas. Indeed, much of what the more expensive infrastructure systems and environmental services do is to break the connections which create local problems: to stop the faeces from entering the drinking water; to stop the waste from blocking the water flows; to prevent the flies from reaching the faeces and the food; and so on. Where the infrastructure and services are lacking, environmental interconnections are central to the creation of local health hazards, and it is difficult to deliniate "water" and "sanitation" as separate concerns.

There is also another sense in which water and sanitation are related to other household and neighbourhood level environmental problems. They follow somewhat similar patterns as one moves from low-income to wealthy neighbourhoods or from poor to wealthy cities. Indeed, before looking at some results, I would like to explain why we came to focus on health-related environmental problems, and at the same time to make some rather crude generalisations about environmental problems and wealth, so as to provide a framework for interpreting the results. To those of you who have always worked in the health field, it may seem self-evident that one should focus on health-related concerns when assessing the urban environment. But the watchword in most discussions of urban environments these days is sustainability. Environmental burdens can cause severe health problems without being unsustainable. Indeed, the problems disadvantaged neighbourhoods face with smoky kitchens, contaminated water, poor sanitation, insect infestation and the like is not that these environmental burdens are ecologically or socially unsustainable, but that they ought not be sustained.

This brings us to an important difference between the primary environmental concerns of the wealthy and those of the poor. To simplify radically, the environmental problems of the urban poor tend to be more localised, more immediate, and more directly threatening to health. The environmental problems of the wealthy tend to be spatially more dispersed, temporally more delayed, and to bear less directly upon health. Thus, the environmental problems of the urban poor are typified by bad water and sanitation and smoky kitchens. The environmental problems of the urban wealthy are typified by global warming and global resource depletion. To an even rougher approximation, the typical problems of the middle income megacities are the intermediate scale problems, such as urban ambient air pollution.

The logic behind this spatial displacement is relatively straightforward. The wealthy use more resources and create more waste, but they also draw on a larger resource base and devote part of their wealth to avoiding exposure to unhealthy or unpleasant pollutants. The end result is that the problems of the urban poor affect predominantly other poor households, while the environmental burden of the wealthy falls on an ever expanding public. Indeed, it is not at all uncommon for environmental problems arising from the narrow pursuit of wealth, to fall predominantly on the poor.

Accompanying the shift in the scale and immediacy of environmental problems, is a shift from issues of health to those of sustainability. Again, there is a relatively straightforward logic to this shift. While the threat of intense environmental insults in and around people's homes is primarily to the health of the inhabitants, the threat of the broader environmental burdens is more likely to involve very indirect routes to undermining human welfare, and take time to manifest itself. Whether these broader environmental burdens are sustainable is often a very pertinent question.

One reason why issues of sustainability have taken such a prominent place in recent environmental debates is probably that they are of particular concern to the wealthy. A century or so ago, when many of the wealthiest cities of the world had such serious sanitary problems that even their wealthy elites were regularly threatened, the precursor of the environmental movement, the sanitary movement, had a clear health focus. Now, the poor and wealthy of the world tend not to live in such close quarters. The spread of infectious diseases is not perceived as such a threat, despite recent rumblings, and a popular new book with the evocative title: *The Coming Plague (Garrett 1994)*. Now environmental awareness is defined not so much in terms of understanding the health implications of local environmental management, but in terms of understanding the contribution of local environmental management to broader issues of sustainability. To the extent that health remains a central aspect of environmental concerns, it is largely in terms of potential future threats. When the Stockholm Environment Institute initiated a study of urban environmental problems in 1991, we decided fairly early on to focus explicitly of the household-level problems. The goal of the study was to be threefold: to support environmental management efforts in each of the cities; to develop methods applicable to other cities; and to inform international discussions on urban environmental issues. In each city, the study covered water, sanitation, air pollution, solid waste disposal, food contamination, pests and crowding. Each major problem area was assessed not only in terms of the physical processes and severity, but also of the health risks involved, the priorities of local residents, and their opinions regarding what needs to be done. Surveys of about one thousand households in each city, and physical testing in a subset of two hundred, provided the core empirical base. In each city the households were selected using a multi-stage design, which conformed to the available sample frames in each city. The selection procedures were designed to provide a representative cross section of households.

Selected Results

Conventional wisdom would have it that Third World mega-cities like Jakarta, and especially São Paulo, are the epitome of urban environmental distress. Yet the household level environmental indicators presented in Table 1, including but not limited to water and sanitation, tell a very different story. For almost every indicator, average conditions improve between Accra and Jakarta, and improve further between Jakarta and São Paulo. The more detailed data clearly support this finding. They would seem to corroborate the view that relative poverty rather than city size which determines, at least to a first approximation, exposure to these household environmental problems. Were the focus on city-level problems, the picture would be quite different, of course. Even per capita, São Paulo has the highest municipal waste generation, the largest quantities of sewerage

Table 1. Household Environment Indicators in Accra, Jakarta and São Paulo - 1991/92

Environment Indicator	**Accra** % of Sample (N=1000)	**Jakarta** % of Sample (N=1055)	**São Paulo** % of Sample (N=1000)
Water No water source at residence	46	13	5
Sanitation Share toilets with > 10 households	48	14-20	<3
Solid Waste No home garbage collection	89	37	5
Indoor Air Main cooking fuel wood or charcoal	76	2	0
Pests Flies observed in kitchen	82	38	17

Source: McGranahan and Songsore, 1994.

released into open water, and the highest concentration of air polluting industries and automobiles. Overall, Accra's city-wide environmental problems pale beside those of both Jakarta and São Paulo. Nevertheless, the environmental health problems faced by Accra's residents are more severe.

Much the same applies within the cities, as indicated in this Table 2. The same household environmental indicators portrayed in Table 1 for the three cities also vary widely across neighborhoods in Accra grouped by wealth. And again, contributions to Accra's city wide problems would look very different - it is the wealthy who use large quantities of water, drive automobiles, and generally live a more resource and waste intensive lifestyle.

Table 2. Selected Household Environment Indicators in Poor, Middle Class and Wealthy Neighbourhoods of the Greater Accra Metropolitan Area - 1991/92

Environment Indicator	**Poor** % of Sample (N=818)	**Middle Class** % of Sample (N=131)	**Wealthy** % of Sample (N=51)
Water No water source at residence	55	15	4
Sanitation Share toilets with > 10 households	60	17	2
Solid Waste No home garbage collection	94	77	55
Indoor Air Main cooking fuel wood or charcoal	85	44	30
Pests Flies observed in kitchen	91	56	18

Source: Stockholm Environment Institute / University of Ghana households survey.

Of course, not all the statistics collected follow this clear pattern. Complexities arise as soon as one starts to look more carefully at the results. Looking across the three cities, for example, samples taken from household water sources (e.g. tap, overhead tank or well) were more likely to show evidence of faecal contamination in Jakarta than either Accra or São Paulo. Of about 200 samples in Jakarta, 60% had positive counts of faecal coliform, as opposed to slightly over 10% of the roughly 130 samples in Accra. However, virtually all of the households in Jakarta boiled their water. The share of samples taken from boiled water in Jakarta indicating contamination was far lower than the share of samples taken from household water containers in Accra, almost 90% of which had positive counts of faecal coliform. Thus, it would seem that water handling and treatment after the households obtain the water decreases the problem in Jakarta, and creates a problem in Accra. In a sense, the correlation between average wealth and average problem severity is reasserted partly because access to water is a more serious problem than water quality at the source, and partly because wealthier households can afford to boil water.

Boiling water is a cultural as well as economic phenomenon. The water is boiled to make tea in Jakarta. This adds another layer of complexity. The extent to which Jakartans drink tea instead of plain water because it is safer is debatable. Indeed, it is part of an ongoing corridor debate on water policy - would the roughly 1% of the cities income devoted to boiling water for tea be saved if the household water supplies were made potable? Anybody who thinks they have an easy answer should think again. On the one hand, tea drinking practice is now central to the etiquette of most Jakartans. On the other hand, Jakartans were once known for not drinking tea, and their relatively high rate of diarrhoeal diseases was ascribed to this (Abeyasekere 1987). Moreover, mineral water has greatly increased in popularity in recent years.

Similar complexities can be found in the other cities. In Accra, there are questions about the different types of water containers. In São Paulo, the first set of water tests seemed to indicate chlorine levels below standard in a very significant share of the 1,000 households. However, further tests revealed that for the most part the decline was occurring in the household water containers, rather than the piped water system. This seemed to explain the anomaly that the low chlorine water was less frequent in poor neighbourhoods - the water samples in poor areas were more likely to be taken directly from the piped supply. The fact that their water was more likely to meet the chlorine standard did not mean the poor households were better off - the lack of a large storage container also means one is more susceptible to supply disruptions, which is likely to have a larger health effect.

There were also some patterns which were similar across certain parts of the different cities - where parts of Jakarta displayed a similar pattern to most of Accra, for example. This applies, for example, to the issue of water pricing.

In the post second world war period, it has been very common for governments to subsidise environmental services, such as water and to a lesser extent sanitation. (Actually, subsidise is a misnomer, since a subsidy is a payment designed to keep prices down, while in many countries the prices have been kept down via direct controls, and truly compensatory payments are often not made.) The economic logic for providing subsidised sanitation and even water in low income areas is very strong. Nevertheless, such strategies have come under a great deal of criticism in recent years from the international donor community, and the situation in both Accra and parts of Jakarta illustrate why.

Essentially, the 'subsidised' services have ended up going to the **relatively** wealthy. Many of the poor who depend on piped water still end up buying water on what amounts to a supply-constrained free market. In Jakarta, they buy from vendors who sell door to door. In Accra, they often buy from people who do have connections. But in both cases, and especially Jakarta, they pay far more than the rate at which the water utility officially sells water. As a result, the poor pay considerably more per litre than the wealthy. This means that, in Jakarta for example, the poor are far more adversely affected by ground water salination.

Figure 1 illustrates the extent to which people must change their drinking water sources when the shallow aquifer is affected by saline intrusion from sea water drawn into the shallow aquifer. Whereas well water dominates for the roughly 60% of the population that lives in areas where saline intrusion is insignificant, piped water connections and private water vendors dominate in the salinated areas.

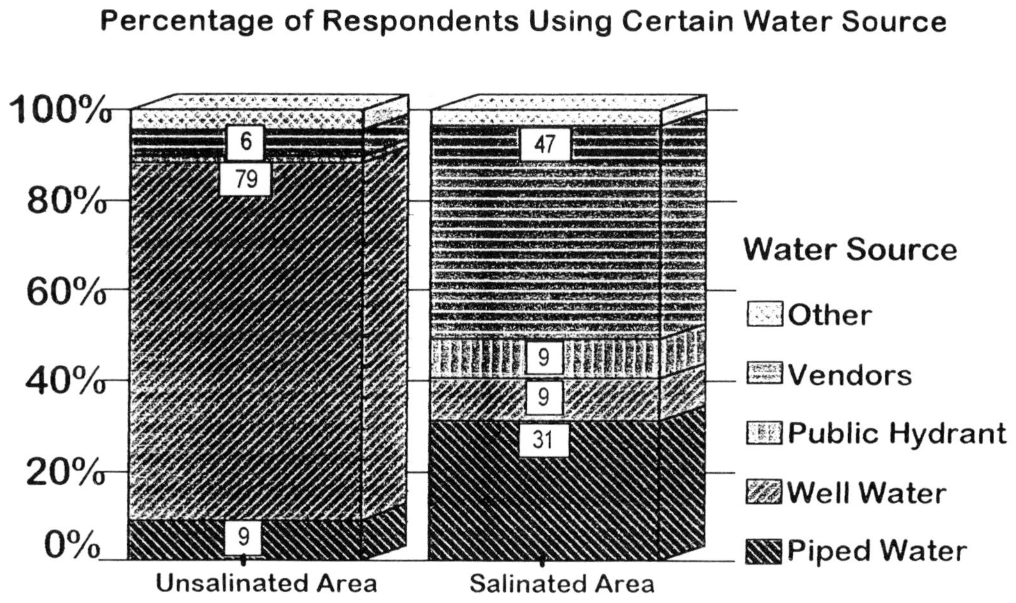

Figure 1. Comparing Principal Drinking Water Sources in Areas with Saline and Non-Saline Groundwater in Jakarta (1991/92)

Source: Kjellen, Marianne, Anna Bratt, and Gordon McGranahan. "Water Supply and Sanitation in Low and Middle Income Cities: Comparing Accra, Jakarta and São Paulo." : Stockholm Environment Institute Draft Working Paper, 1995.

As illustrated in Figure 2, whether a household in a salinated area is more likely to have a piped water connection or to buy from a private vendor depends upon the wealth of that household. Nearly two thirds of the poorest households buy from vendors whereas two thirds of the wealthy have their own connections.

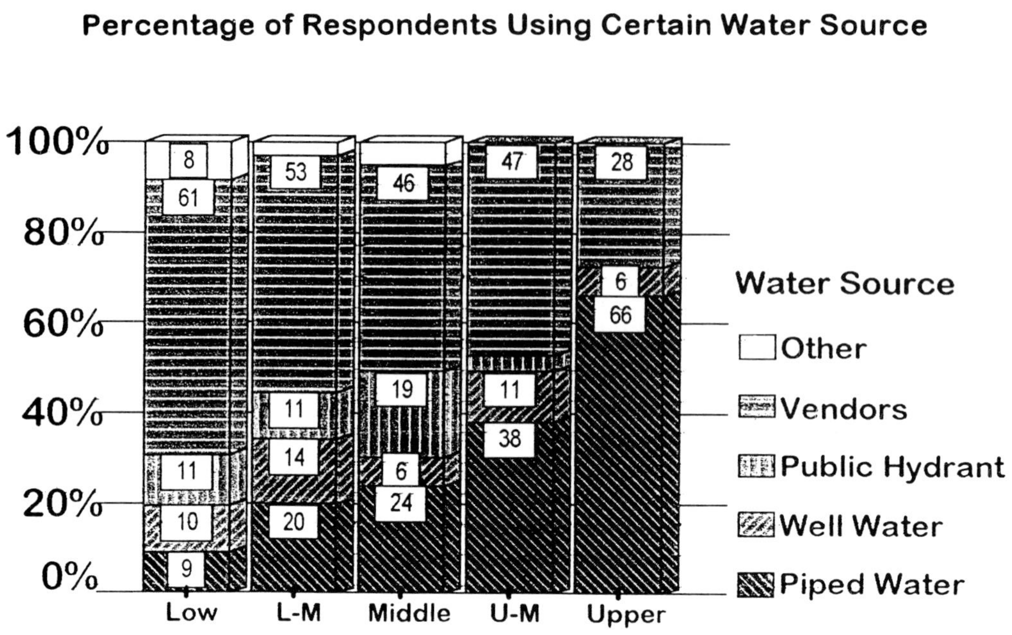

Figure 2. Principal Drinking Water Source in Salinated Areas of Jakarta, by Wealth Quintiles (1991/92).

Source: See Figure 1.

Note 1: The five wealth groups, of roughly equal size, relate to a wealth index based on the households' ownership of consumer durables.

Note 2: Mineral water users are excluded.

The formal tariff structure is such that households consuming little water are charged less per litre than those using large quantities of water. Looking at this tariff alone, one might expect the poor to pay less than the wealthy. However, since the vendors charge far higher prices per litre than the public utility, the poor actually end up paying higher prices, on average, for their drinking water than the wealthy. This is illustrated in Figure 3.

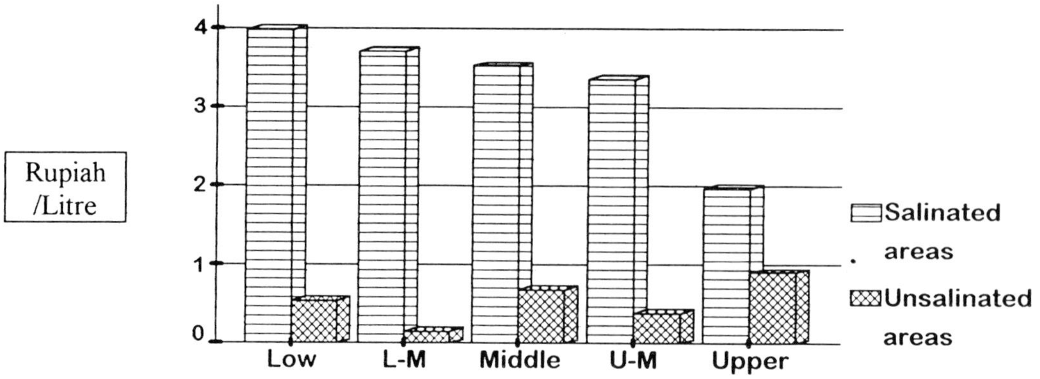

Figure 3. Average Price Paid per Litre of Drinking Water in Jakarta, by Geographical Area and Wealth Group (1991/92).

Source: See Figure 1. Note 1 and 2 in Figure 2 applicable.

The problem with price controls, as they have often been applied, is that if the water is supply constrained, as it seems to be in many parts of Jakarta, the price water vendors charge is largely independent of the utility tariff. Indeed, price controls may actually prevent the utility from being able to expand, and lead to higher prices on the unofficial markets. So one can easily end up with the relatively wealthy subsidised and the very poor effectively being charged more than cost.

Unfortunately, the situation with water pricing is characteristic of a range of situations in which political slippage allows government policies, seemingly designed very reasonably and equitably, become so compromised that they end up no more equitable than many private sector solutions. Despite the high prices they charge, the private vendors are providing a real service for the low income households. Attempting to control the vendor prices would probably just make matters worse.

But the problems which have arisen are not inherent to subsidies. And the criticism that has been mounted on subsidies in recent years provides little indication of how the problems the subsidies **failed** to deal with will be solved. The results from Jakarta show very clearly that there is a big suppressed demand for improved water services among the wealthy (Surjadi et al. 1994). The most obvious strategy for a more **profit** oriented utility would be to invest in improved water services to the wealthy, and ensure that the utility itself gets the high prices from scarce water sales in low income areas. But to deal with the water problems of the poor in a constructive manner, a very different approach is required. Two elements of any such approach are:

1) A more sophisticated and realistic understanding of how the water supply and distribution and use systems operate. This means understanding not only the incentives within the utility, but how the vendor or community based systems interact with the utility, and how the households adapt to the different water supply situations.

2) A means of ensuring that the pressure to improve conditions in the low income neighbourhoods does not end once the formal policy has been defined.

More crudely, it seems clear that the emphasis should be on making it easy and cheap for people to get piped water connections in their home or house compound, assuming they have no other in-house water source, rather than making the water itself cheap. Making water cheap for households with connections is likely to lead to overconsumption and environmental externalities in wealthy households. If it prevents the extension of the supply system in low income areas, it will also lead to underconsumption and environmental health externalities in poor households. On the other hand, cheap and easy water connections are likely to benefit principally those households for whom the environmental health externalities are more serious, and the likelihood of overconsumption is less so.

Turning to sanitation, the technological differences between the cities are very great. In São Paulo, 74% of the toilets were water closets connected to the sewerage system, with most of the remainder either connected to septic tanks (6%) or flowing directly into streams (14%). In Jakarta most people use private aqua privies (67%), while others (about 20%) used communal toilets. In Accra, septic tanks and drained toilets combined only serve 36% of the households, with most people using pit latrines (41%), and others pan latrines (20%) or no sanitation technology at all (4%). There are important technological issues concerning sanitation in each of the cities. On the other hand, the quality of sanitation technology is a bit like the quality of water - it tends to be overemphasised. What came out more clearly as a problem in both Accra and Jakarta was when the supply of sanitary facitlities was inadequate not in its quality but in its quantity. In Jakarta, the shared toilets were clearly in worse condition than the unshared, and not just because the users were poorer. And this is not just a question of public versus private ownership. Shared private toilets were also in poor condition. In Accra, not only were the shared toilets in worse condition, but the people using them were less healthy. Statistically, the sharing of toilets was far more closely associated with diarrhoea than the type of toilet. Indeed, the pan latrines, typically considered the worst technology, and the one being phased out, were associated with a lower diarrhoea prevalence, probably because they were less heavilly shared (Benneh, et al. 1993). More generally, the sanitary conditions facilitating the spread of faecal-oral diseases extend beyond water and sanitation issues narrowly defined. Especially, in Accra, the results also illustrate that problems like childrens' diarrhea reflect a complex set of interconnected environmental problems. It is not very meaningful to assign risks to particular factors in such circumstances.

Summary and Conclusions [2] Although Accra, Jakarta and São Paulo are all Third World cities, their living environments vary radically. By and large, both within and between the cities, poor conditions go hand in hand with economic poverty.

In Accra, the majority is poor and lack proper access to water and sanitation. On average, however, the richer households in Accra are as likely to have access to services as the average Jakartan. Similarly, the wealthier households in Jakarta enjoy an access situation similar to the average São Paulo dweller.

The wealthiest in all three cities are generally connected to a piped public water supply system. But only in São Paulo do most households enjoy water and sanitation services provided directly by the government. In Jakarta, both water and sanitation are most often a household affair, with water from wells and aqua privies for excreta disposal. In Accra, water sources and sanitary facilities are most often shared. These institutional differences are related to the economic wealth of the cities. Accra's widespread sharing is a symptom of poverty, while São Paulo's far-reaching public services reflect wealth.

When it comes to water quality, the situation is not as clearly determined by wealth. Still, the drinking water of the poorest quality is most likely to be ingested in Accra, and those least likely city to ingest faecally polluted drinking water live in São Paulo. The factors influencing water quality however depend on a combination of geographical conditions, water management, and cultural practices.

To facilitate private sanitary facilities seems to be an important part of any strategy to improve public health. An important aspect here is that high levels of sharing of facilities, more so than the use of certain unsophisticated technologies, seem detrimental to hygiene and health. A less sophisticated type of toilet may be better than one shared with many other households.

The geographical conditions also influence water quantities available, at what quality and what price. Where the ground water in Jakarta is not salinated, poorer households can rely on wells for all their water needs, and on average pay less for their water than the wealthy. But in the salinated areas, on the other hand, it is not possible to drink the well water, and on average poor households pay several times more per litre for their drinking water than the wealthy households, who are more likely to have piped water connections.

Where water supply is constrained, low official water prices typically favour only the richest. Furthermore, water utility companies, partly as a consequence of providing water without fully recovering the supply cost, are often unable to extend services, again inordinately disfavouring the poor.

In most developing countries the public sector has largely failed to provide water and sanitation to the poor. The standard solutions, piped water and sewerage systems, typically planned and implemented in a top-down fashion by a government utility, are those which characterise rich cities and neighbourhoods. These standard solutions tend to be too costly and only implemented in certain parts of the cities, notably where the privileged people live.

Still, while running water and water-borne sewage systems may exacerbate city-wide problems, they often reduce household exposure to faecal material. Technically sophisticated systems, when provided to individual households, do not require the households to have a good understanding of environmental issues or to participate in environmental management.

On the other hand, lower-cost solutions generally require much higher environmental awareness and good hygiene practices on the side of the users in order to function well and efficiently eliminate faecal-oral routes of disease transmission. For any intervention, the less comprehensive environmental services provided, and the higher the level of sharing, the more important it is to understand the local context and respond to local priorities and the need for sophisticated management skills at the local level.

This means that for the range of possible solutions to problems in low-income areas, management efforts at the local level are particularly important. With weak local governments it is natural to look towards the private sector, non-governmental organisations and community-based organisations for help. Collective action within neighbourhoods can do much to resolve sanitary problems. Moreover, non-governmental and community-based organisations can help foster a better local awareness of household environmental problems, and place local environmental improvement on the political agenda.

However, community action should not be seen as a panacea. The scope of local organisations is often severely restricted, particularly in poor neighbourhoods. Urban neighbourhoods find it difficult to work collectively at the best of times, and governments often obstruct attempts to organise poor neighbourhoods for fear of the political consequences.

Still, governmental action and information can do much to foster and complement local action and to facilitate market solutions. Regulations on minimum standards, applied constructively can raise the quality of infrastructure and housing. Flexibility is needed: Rules that may serve as an incentive at one level may be prohibitive to poor and underdeveloped areas. Furthermore, for individuals to take action to improve the situation, home ownership is a crucial aspect. Security of tenure, both legally and in practice (as perceived by the residents) is a crucial aspect to be considered before investing in fixed assets.

A strategic approach to solving household environmental problems in poor neighbourhoods would combine the strengths of governments, local organisations, markets and the households themselves. Problems experienced in poor neighbourhoods must be a concern of the society at large. Unfortunately, the present situation is too often one of where the problems of the poor, are suffered by the poor, and dealt with only by the poor. Whereas, on the contrary, the problems of the rich, are suffered by the public, and dealt with by the government.

Notes

1. More details can be found in the published city reports (Benneh, et al. 1993, Jacobi 1995, Surjadi, et al. 1994) or articles (Jacobi 1994, McGranahan and Songsore 1994, Songsore and McGranahan 1993, Surjadi 1993). A more extensive paper on the water and sanitary conditions in these three cities (Kjellen, et al. 1995) can also be obtained directly from the Stockholm Environment Institute (Box 2142, S-103 14, Stockholm, Sweden).
2. These conclusions are from Kjellen, et al. 1995.

References

Abeyasekere, Susan. *Jakarta: A History*. Singapore: Oxford University Press, 1987.

Benneh, George, Jacob Songsore, John S. Nabila, A.T. Amuzu, K.A. Tutu, Yvon Yangyuoru, and Gordon McGranahan. "Environmental Problems and the Urban Household in the Greater Accra Metropolitan Area (GAMA) -Ghana." : Stockholm Environment Institute, 1993.

Garrett, Laurie. *The Coming Plague: Newly Emerging Diseases in a World Out of Balance*. New York: Farrar, Straus and Giroux, 1994.

Jacobi, Pedro. "Environmental Problems Facing the Urban Household in the City of São Paulo, Brazil." : Stockholm Environment Institute, 1995.

Jacobi, Pedro R. "Households and environment in the city of São Paulo: Problems, perceptions and solutions." *Environment and Urbanization* 6, no. 2 (1994): 87-110.

Kjellen, Marianne, Anna Bratt, and Gordon McGranahan. "Water Supply and Sanitation in Low and Middle Income Cities: Comparing Accra, Jakarta and São Paulo." : Stockholm Environment Institute Draft Working Paper, 1995.

McGranahan, Gordon, and Jacob Songsore. "Wealth, health, and the urban household: Weighing environmental burdens in Accra, Jakarta, and São Paulo." *Environment* 36, no. 6 (1994): 4-11, 40-45.

Songsore, Jacob, and Gordon McGranahan. "Environment, wealth and health: Towards an analysis of intra-urban differentials within the Greater Accra Metropolitan Area, Ghana." *Environment and Urbanization* 5,no. 2 (1993): 10-34.

Surjadi, Charles. "Respiratory diseases of mothers and children and environmental factors among households in Jakarta." *Environment and Urbanization* 5, no. 2 (1993): 78-86.

Surjadi, Charles, Lesmana Padhmasutra, Dotti Wahyuningsih, Gordon McGranahan, and Marianne Kjellén. "Household Environmental Problems in Jakarta." : Stockholm Environment Institute, 1994.

Paper 14. SEEKING COMMUNITY-BASED SOLUTIONS TO ENVIRONMENTAL PROBLEMS IN LOW-INCOME URBAN SETTLEMENTS

John Thompson, International Institute for Environment and Development
3 Endsleigh Street, London WC1H ODD, UK

Introduction

There are two persistent myths about participatory, community-based programmes in low-income, inadequately-serviced, urban settlements. First, it has been argued that these initiatives cost more than conventional development programmes. Second, some claim that such programmes take longer to accomplish the same objectives. It is possible that additional costs may be incurred at the outset in strengthening capacity at local level, but these costs are significant only when local organisations have been so eroded that substantial time and resources have to be devoted to building their capacity and competence. Furthermore, even when initial costs of human development and institution building are high, they are more than offset by subsequent gains.

Evidence increasingly indicates that, when the institutional framework is right, participatory programmes actually cost less and are quicker to implement. In many participatory programmes, the typical pattern has been a slow build up period, while time is invested in community organisation and setting the rules for interaction, followed by relatively rapid disbursements. Once the participatory process has been set in motion, the benefits of community-based development include increased efficiency and cost effectiveness. In addition, when the success of projects depends heavily on changes in behaviour at the local level, promoting participation in community-based programmes may be the only means of meeting objectives.

Proponents of participatory research and development approaches claim that they are dynamic and flexible ways of gathering information about, with, and by local people and their own conditions and livelihoods.

As experiences and insights grows, it is evident that local people, who had previously been viewed as passive 'subjects', 'clients' or 'beneficiaries', had much to contribute to the research and development process. As these approaches were adapted and modified further, the depth and validity of local people's experiences and knowledge became clear. Thus, by the late 1980s, much of the attention had shifted from 'rapid' to 'participatory' research and development. Today, these participatory approaches are being applied in a wide range of social and ecological contexts, and are shaping and influencing national, regional and international research and development programmes and policies worldwide (Chambers 1994 a, b, c). Although more commonly associated with rural development and primary health care programmes, there is now a growing body of experience with the use of similar approaches in urban settings.

This paper examines current experiences with the use of participatory methodologies in low-income, inadequately-serviced, urban settlements in solving various environmental problems. Particular attention is given to the application of these approaches for local-level adaptive planning and community-based development by people's organisations and their supporters, rather than on rapid appraisal methodologies used for research and information gathering by external agents.

Participation, Good Governance and Democracy: The Makings of a Civil Society?

Over the past decade, the international donor community and national governments have begun to pay a significant amount of attention to participation, along with related issues like good governance and democracy (World Bank, 1994). In part, this is because they have come to recognise that the modernist models of development they wish to foster will not be embraced and consolidated until effective demand for them can be articulated by the relevant sections of the society. From an optimistic viewpoint, it could be argued that this is because the values of human rights and democracy have been acknowledged as part of the `new' development paradigm and portrayed as something that should not be sacrificed in the name of growth. In a more cynical vein, it could be asserted that donors and states perceive democratic participation as part of a broader process of structural reform, the shift towards market-oriented economies, and the decentralisation and privatisation of state services (Thompson, 1995; Adepoju 1993; Gibbon *et al* 1993; Nelson and Eglinton 1993). From this perspective, participation takes on a more pernicious hue, as it implies that people's participation in development is little more than a hollow, cost-cutting strategy aimed at placing greater responsibility on local communities for their own development activities, while reducing or withdrawing external support and withholding civil liberties.

Some have suggested that if local organisations and community-based approaches are strengthened, by channelling funds to them and by increasing their control over development initiatives, so too will be civil society as a whole. Similarly, programmes and policies that aim to create decision-making processes in which local organisations and associations have a presence allow new means through which civil society can exert influence over public decisions and open that public process to more scrutiny. In this manner, participatory development is directly linked to state accountability, empowerment of local groups, and transparency in decision-making.

Traditional, class-based organisations have had declining membership in many countries and have suffered particularly from the reduction in formal sector employment. Such a trend is directly linked to the promotion of free markets and the decline in public sector employment. It reflects an increase in part-time working, contract workers, and the relocation of industries to low-wage countries.

At the same time, there has been an increase in "social movements" grounded in identities related to gender, place and ethnicity. The apparent strength, ubiquity and diversity of these movements has resulted in a new attention and new literature, although this builds on a previous interest among some authors (Hulme 1994; Farrington and Bebbington 1993; Uphoff 1993, 1992; Korten 1990; Eckstein 1989). This literature has emphasized that movements are characterized by the concern to demand rights of citizenship from the state, and to express social resistance to the many ways in which state and wider society have dominated low-income sectors. These observations generate a vision of civil society as a more complex mosaic of different interest groups than most class analyses had suggested (Scoones and Thompson 1994). At the same time, they are less concerned with revolutionary change than many of the analyses of the 1960s and 1970s (for example, the dependency and Marxist schools). Rather than emphasize the overthrow of the state and the nationalisation of the means of production, they stress the need to make the state more transparent, more open to public scrutiny, and more responsive to societal concerns. This, in turn, leads to the argument that these social movements must be strengthened if this is to be achieved.

The literature also recognizes that the traditional view of the state as powerful institutions with the ability to intervene successfully to support capitalist interests is also flawed. In many cases, the state is weak and ineffectual (Diaz Albertini 1993). Many of these groups have tried to identify a new strategy in relation to the state. Within the old model, groups working with and for low-income communities generally adopted one of two strategies; they either provided welfare services to mitigate the worst effects of poverty or they adopted a political stand and undertook advocacy and political education to organize direct resistance. In response to the new situation, many of those working to support low-income communities began the direct provision of services and material assistance to low-income groups. However, such service provision was often not considered an end in itself but the first step in a complex strategy to develop and recruit support for new state programmes. Through participation in new programmes, communities can begin to realize their needs and learn how to develop successful strategies to realize them.

In developing effective research and development interventions *for* low-income households, many professionals have come to recognize the essential participation of these households, not only in creating the political pressure to secure resources but also in the design of the intervention. Conventional ways of working with communities, with an emphasis on questionnaire surveys and professional style dialogue, have not proved effective and new methods have been needed. In this context, professionals have both sought to use participatory methods developed by others and have experimented with new ways of working with communities.

In Cote d'Ivoire, for example, the shift to local-level maintenance was facilitated by the government water department (SODECI, *Societé de Distribution d'Eau de Côte d'Ivoire)*. Local groups have been responsible for maintaining 13,500 waterpoints and reduced breakdown rates from 50 percent to 11 percent at one-third the cost. The results were sustained only in those villages that had high demand for the rehabilitated water points and where well functioning local organisations already existed (Hino 1993).

The activities of the institutions of civil society - NGOs, residents associations (including those formed by squatters and by tenants), self-help housing associations and cooperatives - have therefore attracted the interest of donors for their role as development organisations. Many have assumed a self-help and developmental role. Many design their own projects and negotiate the support they require directly from the state and donors. These institutions have also attracted interest for their role as political agents that can also help make the state more accountable, especially when groups of these institutions form coalitions or federations, as in the National Federation of Slum Dwellers in India or the South African Homeless People's Federation.

In recent years, many community based organisations, and the agencies that support their efforts have had a crucial role in helping local groups exercise greater control over the development process and articulate their interests and priorities to governments and donors more clearly. The next section describes some of the basic elements commonly associated with these participatory approaches.

What are Participatory Approaches?

Participatory approaches are flexible, process-oriented methodologies combining guiding principles, core concepts and sets of interactive techniques which have been developed in order to better realise high levels of community participation in official development programmes and, more importantly, give local people greater control over the process of development. Participation, as a principle, is now commonly accepted to be a important component of successful development programmes although its meaning is not clear. Pretty (1995) identifies seven different types of participation in development programmes

Table 1. A Typology of Participation

Typology	Characteristics
1. *Passive Participation*	People "participate" by attending meetings where they are told what is going to happen or has already happened in their place. It is a unilateral announcement by an administration or project management without any consultation with community members. The information being shared belongs only to external professionals.
2. *Participation in Information Giving*	People participate by answering questions posed by extractive researchers using questionnaire surveys or similar approaches. People do not have the opportunity to influence proceedings, as the findings of the research are neither shared nor checked for their "trustworthiness".
3. *Participation by Consultation*	People participate by being consulted, and external agents solicit and listen to views. These external agents define both problems and solutions, and may modify these in the light of people's responses. Such a consultative process does not concede any share in decision-making, and professionals are under no obligation to take on board people's views.
4. *Participation for Material Incentives*	People participate by providing resources, such as materials or labour, in return for food, cash or other material incentives. Much on-farm research falls in this category, as farmers provide the fields but are not involved in the experimentation or the process of learning. It is very common to see this called participation, yet people have no stake in prolonging activities when the incentives end.
5. *Functional Participation*	People participate by forming groups to meet predetermined objectives related to the project, which can involve the development or promotion of externally initiated social organisation. Such involvement does not tend to be at early stages of project cycles or planning, but rather after major decisions have been made. These institutions tend to be dependent on external initiators and facilitators, but may become self-dependent.
6. *Interactive Participation*	People participate in joint analysis, which leads to action plans and the formation of new local institutions or the strengthening of existing ones. It tends to involve interdisciplinary methodologies that seek multiple perspectives and make use of systematic and structured learning processes. These groups take control over local decisions, and so people have a stake in maintaining structures or practices.
7. *Self-Mobilisation*	People participate by taking collective actions without the intervention of external agencies. Such self-initiated mobilisation and collective action may or may not challenge existing inequitable distributions of wealth and power.

Source: Pretty 1995.

ranging from "passive participation" when people are simply told what is going to happen to them, to "self-mobilisation" in which local people are active agents of change independent of external organisations (Table 1). The former sees participation as a means to an end, while the latter views participation as an end in itself.

Participatory approaches intend to facilitate the integration of community members into debate and, in some cases, to enhance community control of resource allocation and planning processes. They include a range of different activities designed to:

◊ increase awareness and understanding among the key stakeholders at the local level;

◊ improve the quality and quantity of information about local conditions;

◊ identify viable local development options;

◊ identify local and external resources for such options;

◊ enable local people to identify constraints, set priorities and take action;

◊ strengthen the self-confidence and capacities of local organisations;

◊ develop and support mechanisms to resolve local conflicts.

The text boxes in this section are illustrative of such approaches and describe three different programmes. Box 1 describes Planning For Real, a methodology developed by the Neighbourhoods Initiatives Foundation in the United Kingdom over the past decade. Planning for Real and associated variants are now being used by a range of local groups both within the UK and in a number of other countries, including Cambodia, Colombia, Ecuador, The Gambia, India, Mexico, Mozambique, Sierra Leone, South Africa, Sudan, Tanzania and Zambia (Gibson 1994, Wratten 1994).

Box 1: *Planning for Real*

I think it made people realise just how much they themselves have got to offer, and that nobody is actually ordinary.

Planning for Real is the label for an approach that employs a cluster of interactive techniques and locally-developed materials. In one sense, they are simply a tool-kit that allows people to explore possibilities, sort out options, rank priorities, share out responsibilities and devise a 'Plan of Action', all without having to endure 'talking shops' which drive everyone up the wall, and eventually out of the door. In another, they are a strategy designed to establish common ground between 'Us and Them' as a basis for a combined operation to create a working neighbourhood. This common ground is the local neighbourhood that everyone knows.

The model: The first step is to use this common knowledge to make a three-dimensional model of the neighbourhood. Part of the tool-kit is a *range of suggestion cut-outs*, visual representations, roughly to scale, of what could be done to turn this anonymous dwelling area into a working neighbourhood. There are many possibilities for improvement: drains to remove standing water, public toilets, new waterpoints, play areas, workshops, trees, bus routes, improved housing. Many informal and formal discussions about the model take place; anyone can put any item anywhere, so long as they move no-one else's suggestion.

Residents are not the only people involved. Local officials are invited along to a formal meeting as long as they are willing to agree to a "house rule" to keep their mouths shut until spoken to. Instead of the professionals graciously presenting their own plans for residents' comment, the residents are consulting the professionals, to establish the range of options, the limitations, the possibilities, so that they can reach their own informed conclusions.

Priorities: The next step is to make out cards corresponding to all the items on the model. Each card has a note of the location and the subject. The cards are then laid out on a giant chart divided into three sections, *NOW, SOON, LATER*. With the same freedom of manoeuvre, and the same anonymity if they choose, people can then transfer the cards to what seem the most realistic priorities for action. At once it becomes obvious that there are comparatively few disagreements.

Resource surveys: The residents who promoted the model go round, house-to-house with a cartoon questionnaire (looking as unlike a government form as possible) to find out in each family who is good at what. In parallel with this, some of the 'moving spirits' amongst officials and politicians are digging out resource information from within the local authority and other outside bodies - tracking down people who could help with advice, materials and equipment which might become available, loans or grants that might grease the wheels of local effort.

	What We Can Do Ourselves	What We Can Do With A Little Help	What Someone Else Must Do for Us
NOW			
SOON			
LATER			

Use of the model shows roughly *Where* innovations might take shape. The resource surveys begin to show *Who* might become involved within the community. The *NOW SOON LATER* chart sets out the *When*. This knowledge fuels real decision-making to produce an immediately practicable *Action Plan*.

The development of participatory approaches in urban areas has been encouraged by two factors. They offer better, more comprehensive information faster than conventional research methods and at lower cost. Secondly, information collection has been shown to be one means by which local residents realize and fully appreciate the value of their own knowledge and gain increasing confidence in their capacity to be important agents in development.

Traditional methods of information collection often involved questionnaire surveys that are time-consuming with long gaps between data collection, data analysis and actions based on results. Costs are often high because such surveys are labour intensive and in many development projects such additional funds are not available. In many towns and cities in the South, comprehensive good quality data are not available from censuses or other official sources and to obtain it would be expensive. Amis argues that, in the experience of training Indian government officials, lack of data has been an excuse for lack of action and that this problem needs to be overcome for them to be effective (Amis 1994). A second problem in the use of questionnaires is illustrated in the account in Cernea of a medical anthropologist. The researcher needed information about toilet behaviour, personal hygiene and health beliefs. Although the original strategy was for a sample survey using a precoded questionnaire, the reluctance of respondents to give accurate information resulted in a change of strategy to key informant interviewing and participant observation (Cernea 1992).

A more fundamental criticism is that such "objective", positivist data collection processes misrepresent and over-simplify the interactive nature of the research "encounter" between external researchers and local people, a process which is highly subjective, complex and value laden. As a result, these procedures have failed to reveal the dynamism of everyday knowledge, the social construction of changing relations of power, or the "life-worlds" of local actors (Habermas, 1987, 1984). Furthermore, they have misrepresented the way in which outside researchers and local people "co-produce" rather than simply "discover" the worlds of their research (Nencel and Pels, 1991; Steier 1991).

Such criticism was first responded to by the development of the interpretavist tradition in sociology with micro-level studies, and then by reflexive ethnographic writings in anthropology (Scoones and Thompson 1994; Clifford and Marcus 1988; Marcus and Fischer 1986). These key debates point to the Western ethnocentrism inherent in concepts about culture, local people's knowledge and the role of outside "agents of change". They show how subjects of social science study have frequently served to reproduce rather than redress race, gender, and other inequalities. Moreover, they indicate that researchers must become part of the investigation within a process that explicitly tackles the complexities of multiple realities.

More recent methodological approaches have extended this debate further by portraying research as a collaborative process, researching *with* and *for* people rather than *on* people (Brown 1994; Sayer 1992; Ricoeur 1981).

In order to address critics who argue that participatory approaches are too subjective, a range of validation techniques have been developed. For example, "triangulation" is the cross-checking of information either by varying the methods, varying the sources of information or varying the investigators. Interdisciplinary teamwork further strengthens this process, as various disciplinary perspectives are brought together in a more holistic framework to analyze and interpret local issues and experiences. In this way, the "trustworthiness" of the data is assured through repeated participant and peer checks and reviews (Lincoln and Guba 1985).

Without community involvement, well intentioned assistance may be wasted or "captured" by higher-income groups, often resulting in a process whereby those the assistance is intended to help are further impoverished (Hardoy and Satterthwaite 1989). With such information, community members can individually and collectively identify and recognize their skills, and also their needs and priorities.

In some cases, participatory approaches have been combined with the creation and/or strengthening of formal community organisations and federations of such organisations. In such circumstances, participatory approaches do more than simply increase the effectiveness of the work of the organisations that use them, they also create a possibility for new alliances and coalitions in order to more effectively realize common objectives.

The Development of Participatory Approaches

The historical development of participatory approaches has largely been within two sectors, agriculture and public health (*PLA Notes*, 1988-present; Scrimshaw, 1992). While they have separate traditions, there is some acknowledgement of their related origins. Within agriculture and rural development, these approaches are associated with the term "Rapid Rural Appraisal" (RRA) and within the public health sector "Rapid Assessment Procedures" (RAP). Since the late 1980s, practitioners of RRA and other rapid methodologies have drawn on other methodological and theoretical traditions and developed a range of participatory approaches, including Participatory Rural Appraisal (PRA). Chambers (1994a) argues that these influences include s participatory research and participatory action research approaches developed from the work of the Brazilian pedagogue Paulo Freire, agroecosystem analysis, applied anthropology and field research on farming systems.

Participatory Rural Appraisal uses similar methods to rapid rural appraisal but, in addition to obtaining the information, is also concerned with enhancing local people's *active* participation in the research and development process. It builds on people's innate visual literacy by employing a variety of diagramming and visualisation methods that enable both literate and non-literate persons to participate actively.

Within the public health sector, interactive approaches can be traced to Rapid Assessment Procedures (RAP), which emerged out of anthropological methods in the late 1980s. These were adapted into tools and procedures that could be used to rapidly obtain information on household and community health and nutrition related behaviour (Scrimshaw and Gleason 1992). While RAP has been used primarily by researchers, the SARAR approach has been designed specifically for development practitioners. SARAR integrates five elements - self-esteem, associative strengths, resourcefulness, action planning, and responsibility - into a participatory framework designed to generate discussion among community members and involve them in identifying health and hygiene problems and solutions. SARAR has been developed and promoted widely by UNDP/World Bank's PROWWESS programme for use in the rural water supply and sanitation sector (World Bank 1994).

Participatory Approaches in an Urban Context

In an urban context, there has been little explicit, widespread development of such methodologies. Although there are an increasing number of case studies of urban community development, in general, very little emphasis has been placed on methodology. Perhaps one of the best known examples is approach taken by the Million Houses Programme profiled in Box 2 below in which the specific methodology of community involvement has always played a clear role. More recently there has been an increased interest in drawing on methodologies developed within a rural context for urban work. This interest includes a broad spectrum of agencies and areas of work including experimentation with the use of rapid assessment procedures for urban poverty and programmes assessments (Leach 1994; Norton 1994) and the incorporation of PALM tools within the UK Overseas Development Administration's Indian Slum Improvement Programme (Francis 1994).

It is also evident that there has been a widespread experimentation with different participatory approaches by groups working in urban areas (*RRA Notes* 21, 1994). In many aspects, similar tools and techniques have been developed based around the use of diagrams, maps and pictures to replace written descriptions. Drawing from both urban and rural experiences, specific tools and techniques used in urban areas are:

◊ *Participatory mapping of the settlement* including the plotting of important landmarks such as rivers, roads, religious centres (mosques, churches, temples, etc.), and public services plus indications of features such as topography. Such maps may be drawn in any open space within the settlement by a group of inhabitants and local informants.

◊ Community members undertaking *household surveys* of the settlement to collect socio-economic data such as numbers of children or resources such as skills of household members.

◊ *Collective modelling of new housing designs* that will better meet the needs of residents. House models can be made from any easily available plastic and cardboard. Models can then be discussed and revised with different groups in the community in order to identify the preferred model.

◊ *Collective planning of new settlement designs* that will better meet the needs of residents. Once a plan of the existing settlement has been made, residents can discuss how to change buildings, reblock plots and/or improve services. Once done, residents can explore how they can move from where they are presently to where they want to be.

◊ *Collective identification of resources* including access, management and control, and including sources of income, health and links to rural areas. For example, small groups can draw pie charts to indicate the share of the community working in different jobs.

◊ *Transect walks,* group walks through a settlement - for example, to identify the different informal sector activities taking place or to identify housing conditions and informally interview local people about issues of concern.

◊ *Seasonal calendars* to identify seasonally occurring events such as illness, availability of employment, dietary patterns and income and expenditure.

◊ *Wealth-ranking and well-being analysis* of all households in the settlement either through households being ranked from first to last or using relative colours to indicate different levels of wealth and well-being based on locally defined criteria.

◊ An historical understanding can be developed either through *trend analysis or life histories*. In trend analysis, discussions with old people are used to plot the provision of basic services in the settlement or to consider how factors such as population and social customs have changed over time. Through small group discussions with accounts of individual life histories, critical events in people's lives and in the life of the settlement can be identified.

◊ *Institutional analysis* of relations with other local and external groups and organisations can be identified through *Venn diagramming*. Different sized circles are given to small groups of participants who then use the circles to illustrate the relative importance of other groups to the community. The size indicates relative their importance; the distance indicates the frequency of their interactions with the community.

◊ *Matrix scoring and ranking* methods are used to identify, analyze and compare various resources (eg., public services) and development options (eg., latrine design, water supplies), and identify priorities.

◊ *Social dramas and role plays* are acted out by local people to provoke discussion about opportunities and constraints facing community members.

◊ Establishment of *formal and informal groups* in the community that can provide a focus for and maintain the momentum of community driven development.

One of the most comprehensively documented examples of the use of participatory methods in urban development is community action planning in Sri Lanka (Box 2 on next page). The objective of the approach is to motivate and mobilise the population of an urban low-income settlement to take the lead in planning and implementing an improvement programme.

Despite relatively little exchange about the use of such methodologies within an urban context, it is evident from a recent review (Mitlin and Thompson 1994) that they are used by groups working in settlements in urban areas to consider a wide range of issues including neighbourhood and community development.

The experiences reflected in the text boxes and in other examples (*RRA Notes* 21, 1994) are that communities have an important contribution to make to improvement programmes if such development is to transform their lives. In the case of Planning for Real, originally developed for use in the UK, this exchange has been made more focused by a Development Planning for Real for use by groups in the South (Wratten 1994).

As urban development has officially become the job of architects, engineers and planners, so small-scale household building processes (whether self-build or by contractor) have been excluded from professional consideration except as "eye-sores" to be declared illegal and to be later replaced (Lankatilleke 1995). Examples from Africa, Asia and Latin America suggest that there is a rich tradition of community-based urban development efforts (Arrossi, et al. 1993; Turner 1988). Some of the better known examples include:

◊ In Fortaleza, Brazil, NGOs have worked with low-income community groups to collectively redesign houses and settlements.

◊ In Manila, the Philippines, women have been exploring critical events in the development of settlement through sharing life histories.

◊ In India, participatory methods have been used to assist in identifying appropriate responses to the earthquake in Maharashtra

Box 2: ***Community Action Planning in Sri Lanka***

"The real authors of the methodology are the community and the officials who worked alongside one another" (Lankatilleke, 1989).

The vehicle for community action planning and management is the interaction/partnership workshop. At such workshops, community members interact as partners with the staff of the National Housing Development Authority, the local authority and the non-governmental organisations. They discuss the problems of the community, identify solutions and formulate plans of action. The community takes responsibility for implementing these action plans in collaboration with the NHDA and other organisations, and to maintain and manage the built environment after the completion of the project.

The first step is a two-day community action planning workshop with participation from all relevant groups. It provides an opportunity for the community to obtain a comprehensive view of its socio-economic situation and to identify its main concerns and priorities. It also exposes the community to the opportunities available for the improvement of its living conditions, as well as the constraints and obstacles that need to be overcome. The participants at the workshop first identify community problems and consider their causes. Possible solutions are then explored and strategies identified. The next stages involve developing concrete action plans and systems to monitor their implementation. Finally, the workshop consider how to ensure that everyone concerned is aware of the plan. Further half-day workshops then consider specific problems and issues.

Settlement Regularisation: In the community action planning approach, the individuals and community play a central role in settlement regularisation. A community workshop determines the broad principles within which the regularisation process should take place, such as the width of roads and footpaths.

The decisions of the workshop on the principles and guidelines are distributed to all households in the settlement. Community leaders inform clusters of households of the day the "blocking-out exercise" will be conducted in their cluster and request the households to be at home on that day. The action planning team visits the cluster to discuss the plot boundaries with each of the households. The team meets with the families in each block to discuss the size of the area and whether or not it can accommodate all the households and, if not, how the problem will be dealt with. As soon as there is an agreement, plot markers are placed to allow all involved to see the implications of the decisions. This will often lead to objections and further negotiations by the affected families. The process of negotiation between the families is the most important part of the exercise. The role of the officials is to ensure that no household can grab more land than has been agreed on by the workshop (which establishes a maximum size for plots in the settlement). In the process, all land disputes are settled on the spot and finally consensus is reached about the re-blocking of the land in the settlement.

Community Building Guidelines. Once land tenure has been regularised, the residents of the low-income settlement are usually eager to start the construction or improvement of their houses. The Urban Development Authority has made a provision in its laws concerning planning and building standards to allow reduced standards in these settlements. In the community action planning approach, representatives of the various interest groups in the settlement work together with health and planning professionals to formulate building codes specific to that settlement.

◊ In Chile, houses have been designed by non-specialists using house modelling exercises.

◊ In Pakistan, the Orangi Pilot Project makes rapid and low-cost surveys of areas that are to be provided with secondary drains by drawing on the community's expertise.

◊ In Zambia, PRA methodologies have been used to identify appropriate donor support for income generation projects in Lusaka.

◊ In Guinea, FAO have developed a programme of support for artisanal fisheries both improving local opportunities and changing national policies.

◊ In Birmingham, England, participatory tools have helped to initiate discussions and development programmes with Bangladeshi immigrants about waste management.

In a joint initiative between NGOs and community-based organisations in South Africa and India, community exchanges have been combined with participatory methodologies in order to "root" the learning process even more solidly within the communities and to rapidly accelerate the capacity of community leaders and members to adopt new roles within the development process (Bolnick and Patel 1994). This partnership is between the People's Dialogue in South Africa together with the recently formed South African Homeless People's Federation and a group of three organisations in India: SPARC, the National Slum Dwellers Federation and *Mahila Milan*, a federation of women's collectives (Bolnick 1993; D'Cruz and Patel 1994). The approaches that are used are described in Box 3 (next page).

This partnership has meant that the People's Dialogue members have been able to benefit from community-based shelter training programmes in India where the National Slum Dwellers Federation an *Mahila Milan* have been engaged in such programmes for over eight years.

The India-South Africa exchange programme is unusual because community leaders themselves have been responsible for demonstrating the value of their respective methodologies and, therefore, for spreading those approach to members within their organisations. A more common strategy has been the use of trained professionals to initiate activities at a community level.

Experiences with Planning for Real suggest that participatory approaches can transform relations between communities and government officials. In place of hostile relations with low community involvement in government programmes, communities that developed their own agendas through Planning for Real drew enthusiastically on government expertise once the official recognised that they should not longer control the development process (Gibson 1994).

Within the community planing approach in Sri Lanka, additional support is available from government staff if required by the community. For example, if the community leaders and the staff of the local authority feel the need for a conceptual layout plan, the Urban Housing Division will prepare such a plan before the on-site blocking-out starts.

Conclusions

Throughout the world, participatory research and development approaches are being applied with increasing frequency in urban contexts with varying degrees of success. By strengthening the process of local decision-making, these approaches have resulted in local people being:

◊ aware of their knowledge and capacities, and gaining self-respect;

◊ able to negotiate as equals with government and other agencies;

◊ capable of devising and initiating strategies to improve their situation;

◊ able to respond to and resolve conflicts within the community.

Box 3: ***The India-South Africa Community Exchange***

The experience-based learning of the training process has two separate but interrelated purposes. First, it enables low-income people to evolve their own understanding of their social and economic environment, not just on a micro-level but via exchange in regional and global arenas. Second, it equips the participants, impoverished residents of informal settlements, with the ability to carry out and drive their own experiential learning programmes.

Shack Counting: Normally the training begins with the physical counting and mapping of all houses and other structures in the settlement and this shack counting exercise always starts with a huge celebration which might be a concert, community based drama, visiting dignitaries or other homeless communities. A rough map of the settlement, drawn a few days before, and a series of photographs of the settlement are displayed. One member of the "training team" is assigned to each section. They become the leaders of groups of people who will assist in counting all the structures in the sections. The informal exchanges that take place during this process are the very soul of the process. This simple process of dialogue and exchange only occurs when the people from communities do the counting. The informal discussions that accompany community driven enumerations are both an outstanding method of mobilisation and an exceptionally accurate way of identifying issues that people in the community regard as relevant. The process helps identify and release the real feelings, frustrations and expectations of oppressed people. The way a squatter responds to the inquiries of a fellow squatter is very different from, and more relevant than, the way that same squatter responds to the social scientist or researcher. In tandem with the shack count and numbering, the training groups draw rough maps of the settlement.

Mapping: As the groups progress, they draw a simple map of the streets and structures. Key landmarks are included, such as drains, sewers, electric lights, rivers and other major features. Once the shack counting and mapping have been completed, the sections are combined into one by a community member who draws well. Once people have demarcated their settlement themselves, they go on to examine landownership and related matters in terms of their own needs and experience.

Surveys: Although this training programme does not follow any set formula, a survey usually follows a shack counting exercise to collect a range of data needed for the development process. Once the information generated by the shack count has been reflected back to the community via a mass meeting accompanied by the graphical display of all the information gathered, the survey is started.

House Models: Members of the host community are encouraged by the training team to imagine the house they would like to live in, and to put that dream on paper. By drawing the house of their dreams, people begin to visualise possibilities in terms of their abilities and their levels of affordability. The houses of people's imaginings are usually too elaborate and costly for their resources. Once individuals have drawn their dream houses, they discuss and revise their houses in groups. The group builds a cardboard model of the chosen house and costs their design. A house modelling competition is held in the community. The chosen design is built using cloth or paper as material. People get together to officially open and view the model house.

Housing Savings Groups: Housing savings groups are loosely structured organisations that enable homeless people to develop financial systems that they control and manage themselves. People will point out that they are homeless and landless because they cannot afford formal housing, they will start to think how they can harness resources so that formal housing is possible.

From experiences to date, a number of questions have emerged about the use of participatory approaches in urban areas. First, what is the appropriate time-frame necessary for ensuring active local involvement and establishing sustainable development processes? Second, how truly participatory are these approaches? Finally, how can locally driven, people centred development programmes link with higher level planning structures and policy-making procedures?

As rapid assessment has moved towards and merged with participatory development, speed of execution has lost much of its importance. While some of the methods may result in the relatively quick gathering and analysis of information, the establishment of sustainable development programmes and procedures is a slow, arduous, time-consuming process. For example, participatory approaches were used in an FAO-assisted programme to develop viable artisanal fisheries organisations in the port areas of Guinea Conakry, West Africa (Reusen and Johnson, 1994). Although certain valuable information was gathered quickly, the actual process required several years because of the need to train government officials, establish and strengthen representative local organisations, and build a core cadre of skilled facilitators capable of supporting local planning and development efforts. Some participatory approaches and methods are used regularly in rapid reconnaissance, emergency relief and poverty assessment activities, but there is no escaping the fact that real bottom-up development requires a long-term commitment.

Community-based development is a complex process and reservations have been expressed about the nature of participation within such programmes. Who actually takes part and who is excluded? Whose voice is heard and whose is silenced or ignored? Rarely, if ever, will 100 percent of the local population take part in all aspects of development. Mosse (1994), drawing on experience in rural areas, suggests that the application of participatory approaches cannot be quick if all local people are to understand and accept the external facilitators interacting with them. Furthermore, community participation, he argues, is likely to favour stronger groups in the community who can represent their private interests as public concerns. In order to ensure that more marginal groups are involved and empowered, significant effort must be invested in consulting widely, creating non-threatening opportunities in which less powerful individuals are given the opportunity to express opinions and challenge prevailing views. There are no shortcuts to getting this process right; it takes time, and a tremendous amount of sensitivity and understanding of local social relations.

How can local planning within the community be linked to higher-level planning structures and policy-making? As community capacity to plan local settlements and projects increases, there is a need to ensure that municipal and city officials openly support and actively encourage such developments. Frequently, autonomous and community driven development initiatives may conflict with official rules and regulations. At a city level, participation is likely to be a more formal process and there is a need to address how local governments can develop planning procedures that are more flexible and responsive to community based initiatives. This will require fundamental changes to the internal procedures and management strategies of the government agencies themselves. New working rules, financial management practices, reporting systems and supervisory methods will be needed. Functions and objectives will have to be clarified before new structures can be designed. At the same time, there must be a shift away from the standardized procedures and specialized units responsible for discrete stages in the development process, and more emphasis must be placed on interdisciplinary sharing and learning. Finally, a range of incentives for reorienting and restructuring systems and structures will need to be developed for rewarding those who facilitate institutional change and promote participatory approaches.

Despite these concerns, there is evidence of growing interest in the use of participatory approaches for neighbourhood improvement programmes in both urban and rural areas. The experiences described here demonstrate a new direction for participatory research and development methodologies that are being used by government, NGOs and community-based organisations. Both the scale of present interest and their widespread acceptance in rural development suggest that their application will continue to grow in urban areas. The challenge ahead is to ensure that with the adoption and use of these participatory approaches will come new institutions and new policies capable of supporting and sustaining the process.

References

Adepoju, Aderanti. 1993. *The Impact of Structural Adjustment on the Population of Africa: The Implications for Education, Health and Employment.* London: James Currey.

Amis, Philip. 1994. Urban management training, action learning and rapid analysis methods, *RRA Notes* 21: 13-16. London: Sustainable Agriculture Programme, International Institute for Environment and Development.

Arrossi, Silvina, Felix Bombarolo, Luis Coscio, Jorge E. Hardoy, Diana Mitlin and David Satterthwaite. 1993. *Funding Community Initiatives.* London: Earthscan Publications Ltd.

Bolnick, Joel. 1993. The People's Dialogue on land and shelter: community driven networking in South Africa's informal settlements. *Environment and Urbanization* 5 (1).

Bolnick, Joel and Sheela Patel. 1994. *Regaining Knowledge: An Appeal to Abandon Illusions.* Cape Town and Bombay: People's Dialogue and SPARC.

Brown, Richard Harvey. 1994. *Social Science as Civic Discourse: Essays on the Invention, Legitimation, and Uses of Social Theory.* Chicago: University of Chicago Press.

Campbell, Andrew. 1994. *Landcare: Communities Shaping the Land and the Future.* St. Leonards, Australia: Allen and Unwin.

Cernea, Michael M. 1992. Re-tooling in applied social investigation for development planning: some methodological issues. In Nevin S. Scrimshaw and Gary R. Gleason (eds.), *RAP: Rapid Assessment Procedures, Qualitative Methodologies for Planning and Evaluation of Health Related Programmes.* International Nutrition Foundation for Development Countries.

Chambers, Robert. 1994a. The origins and practice of Participatory Rural Appraisal. *World Development* 22 (7): 953-969.

Chambers, Robert. 1994b. Participatory Rural Appraisal (PRA): analysis of experience. *World Development* 22 (9): 1253-1268.

Chambers, Robert. 1994c. Participatory Rural Appraisal (PRA): challenges, potentials and paradigm. *World Development* 22 (10): 1437-1454.

Chambers, Robert, Arnold Pacey, and Lori Ann Thrupp (eds). 1989. *Farmer First: Farmer Innovation and Agricultural Research.* London: Intermediate Technology Publications.

Clifford, James, and George E. Marcus. 1988. *Writing Culture: The Poetics and Politics of Ethnography.* Berkeley: University of California Press.

Cornwall, Andrea, Irene Guijt, and Alice Welbourn. 1994. Acknowledging process: challenges for agricultural research and extension methodology. In Ian Scoones and John Thompson (eds) *Beyond Farmer First: Rural People's Knowledge, Agricultural Research and Extension Practice.* London: Intermediate Technology Publications: 98-117.

D'Cruz, Celine and Sheela Patel. 1993. The *Mahila Milan* crisis credit scheme: from a seed to a tree", *Environment and Urbanization* 5 (1).

Diaz Albertini, Javier. 1993. "Non-profit advocacy in weakly institutionalized political systems: the case of NDGOs in Lima Peru", *Non-profit and Voluntary Sector Quarterly* 27 (4) Winter.

Eckstein, Susan (ed.) 1989. *Power and Popular Protest: Latin American Social Movements*, Berkeley: University of California.

Edwards, Michael and David Hulme (eds). 1992. *Making a Difference: NGOs and Development in a Changing World.* London: Earthscan Publications.

Fals-Borda, Orlando and Muhammad Anisur Rahman (eds). 1991. *Action and Knowledge: Breaking the Monopoly with Participatory Action-Research.* London: Intermediate Technology Publications.

Farrington, John and Anthony Bebbington, with Kate Wellard and David L. Lewis. 1994. *Reluctant Partners: Non-Governmental Organisations, The State and Sustainable Agricultural Development.* London: Routledge.

Francis, Sheelu. 1994. PALM in Slum Improvement Projects: a training experience from India. *RRA Notes* 21: 36-40. London: Sustainable Agriculture Programme, International Institute for Environment and Development.

Gibbon, Peter, Kjell J. Havnekik, and Kenneth Hermele. 1995. *A Blighted Harvest: The World Bank and African Agriculture in the 1980s.* Trenton, NJ: Africa World Press.

Gibson, Tony. 1994. *Showing* what you mean (not just talking about it). *RRA Notes* 21: 41-7. London: Sustainable Agriculture Programme, International Institute for Environment and Development.

Habermas, Jürgen. 1987. *The Theory of Communicative Action: Critique of Functionalist Reason*, Vol. II. Boston: Beacon Press.

Habermas, Jürgen. 1984. *The Theory of Communicative Action: Reason and the Rationalization of Society*, Vol. I. Translated by T. McCarthy. Boston: Beacon Press.

Hardoy, Jorge E. and David Satterthwaite. 1989. *Squatter Citizen.* London: Earthscan Publications.

Hulme, David. 1994. Social development research and the third sector: NGOs as users and subjects of social inquiry. In David Booth (ed) *Rethinking Social Development: Theory, Research and Practice.* Harlow, UK: Longman Scientific and Technical: 251-275.

Korten, David C. 1990. *Getting to the 21st Century: Voluntary Action and the Global Agenda.* West Hartford, CT: Kumarian Press.

Lankatilleke, Lalith. 1995. Build Together: national housing programme of Namibia. Paper presented at the International Workshop on Participatory Approaches and Methods in Urban Areas, January 1995, London, International Institute for Environment and Development.

Leach, Martin. 1994. Targeting aid to the poorest in urban ethiopia - is it possible? Rapid Urban Appraisal. *RRA Notes* 21: 48-54. London: Sustainable Agriculture Programme, International Institute for Environment and Development.

Lincoln, Yvonna S. and Egon G. Guba. 1985. *Naturalistic Inquiry.* Newbury, CA: Sage Publications.

Marcus, George E. and Michael M.J. Fischer. 1986. *Anthropology as Cultural Critique: An Experimental Moment in the Human Sciences.* Chicago: University of Chicago Press.

Mosse, D. 1994. Authority, gender and knowledge: theoretical reflections on the practice of Participatory Rural Appraisal. *Development and Change* 25 (3): 497-526.

Nelson, Joan M. and Stephanie J. Eglinton. 1993. *Global Goals, Contentious Means: Issues of Multiple Aid Conditionality.* Washington, D.C.: Overseas Development Council.

Nencel, Lorraine and Peter Pels (eds). 1991. *Constructing Knowledge: Authority and Critique in Social Science.* London: Sage Publications.

Norton, Andy. 1994. Observations on urban applications of PRA methods from Ghana and Zambia: Participatory Poverty Assessments. *RRA Notes* 21: 55-6. London: Sustainable Agriculture Programme, International Institute for Environment and Development.

PLA Notes. 1988-present. Formerly *RRA Notes.* London: Sustainable Agriculture Programme, International Institute for Environment and Development.

Pretty, Jules N. 1995. Participatory learning for sustainable development. *World Development* 23 (8): 1247-63.

Ricoeur, Paul. 1981. *Hermeneutics and the Human Sciences.* Edited and translated by John B. Thompson. Cambridge: Cambridge University Press.

RRA Notes 21. November 1994. *Special Issue on Participatory Tools and Techniques for Urban Areas.* London: Sustainable Agriculture Programme, International Institute for Environment and Development.

Sayer, Andrew. 1992. *Method in Social Science.* 2nd Edition. London: Routledge.

Scoones, Ian and John Thompson (eds). 1994. *Beyond Farmer First: Rural People's Knowledge, Agricultural Research and Extension Practice.* London: Intermediate Technology Publications.

Scrimshaw, Nevin S. and Gary R. Gleason (eds). 1992. *RAP: Rapid Assessment Procedures, Qualitative Methodologies for Planning and Evaluation of Health Related Programmes*. International Nutrition Foundation for Development Countries.

Steier, Frederick (ed). 1991. *Research and Reflexivity*. London: Sage Publications.

Thompson, John. 1995. Participatory approaches in government bureaucracies: facilitating the process of institutional change. *World Development* 23 (9): 1521-54.

Turner, Bertha (ed). 1988. *Building Community*. London: Building Community Books.

Uphoff, Norman. 1993. Grassroots organizations and NGOs in rural development: opportunities with diminishing states and expanding markets. *World Development* 21 (4): 607-622.

Uphoff, Norman. 1992. Local institutions and participation for sustainable development. *Gatekeeper Series*, SA31. London: Sustainable Agriculture Programme, International Institute for Environment and Development.

World Bank. 1994. *The World Bank and Participation: Report on the Learning Group on Participatory Development. Fourth Draft (revised).* Washington, D.C.: World Bank.

Wratten, Ellen. 1994. Bottom-up planning for urban development: The *Development Planning for Real* Pilot Project. *RRA Notes* 21: 83-90. London: Sustainable Agriculture Programme, International Institute for Environment and Development.

Paper 15. LEGAL ASPECTS OF SAFE WATER ENVIRONMENTS IN PERIURBAN KENYA

Kenneth Orie
School of Environmental Studies, Moi University, P.O.Box 3900, Eldoret, Kenya

Introduction

Water permeates the entire fabric of social and economic activities of a society. It is central to the existence of humans, animals and plants. But to sustain this all important role of water, the resource must be safe and useable both in terms of quality and quantity. One may regard safe water as only that which is still in a pristine state. But such water hardly exists, particularly in urban or periurban areas. Yet by the use of treatment plants or clean-up equipment polluted water can be made safe for human, animal or plant use. Functionally, therefore, one can say that safe water is the physical, chemical and biological conditions of water and its environment which do not pose health hazards to humans, animals or plants regardless of whether the water is pristine, treated or cleaned up.

Law does not apply in vacuum. It applies to a given subject and knowledge of that subject is important for the formulation of appropriate law to achieve the desired result in terms of management and compliance. Thus an effective law must capture both scientific facts and values of the people whose interest it is intended to serve. The central issue therefore is the extent, if any, the Kenyan law (legislation) captures scientific facts and customary values about safe water environment, that is, how effective it is as an instrument of ensuring safe water environment and the direction any improvement should follow.

Water quality and quantity can be polluted and diminished respectively by natural forces such as mineralization, salinization or drought. But the major threat to water quantity and quality (safe water environment) is human. Legal control of human behaviour is therefore a crucial element in any comprehensive approach towards ensuring safe water environment whether in periurban areas or otherwise. This approach captures factors like water supply and use, sanitation, disposal of wastes, planning of human settlement with good drainage systems, poverty, health education and culture. Any law or policy designed for safe water environment should take these factors into account. The law of safe water environment applies generically to all parts of the country and therefore it is largely idle to discuss a separate legal regime for periurban areas.

Water Supply, Use and Conservation (Water Quantity)

Four-fifths of Kenya is arid and semi-arid and has been affected by drought in the past (1). There are only eight rivers of permanent flow in the country. The others have seasonal flow, disappearing during dry season. Rainfall is sporadic, favouring a limited area of the country. Only about 11% of the rainfall exceeds 1000 mm (2). and water demand is expected to double and, in certain cases, triple by the year 2010 (3). Burgeoning population, expanding demand for water and increasing water pollution are posing a considerable threat to the limited water supply in Kenya. For example, the municipal council of Eldoret has admitted inability to meet the water demand in the municipality partly because of rapid increase of population, limited water supply and poor operation and maintenance of water supply points (4). The government (5) has acknowledged its inability, due to inadequate manpower and financial resources, to meet its national goal of providing piped potable water to all Kenyans by the year 2000.

Water supply. Roof catchments, shallow wells, springs and rivers could be important sources of water supply, particularly where pipe-borne water is not available. Under the Water Act, the government (the Water Apportionment Board (WAB)) has the responsibility of allocating water to users and regulating the construction of water works through a permit system (6). Under section 28(1) of the Act, it can prescribe the construction of a controlling or measuring device to ensure that a water abstractor does not use, abstract, divert or store more quantity of water than he is authorized. Also, it has the power to reallocate water equitably during drought or in the case of a small watercourse, as it sees fit, notwithstanding a person's right under a permit. Permits can also be varied, altered or revised during water shortage or interference with domestic use or other emergency (7), or where hydrographic survey so indicates (8). Water use permits can be cancelled under s. 99(a)(b) where water is not being applied to beneficial uses.

Water supply in urban and periurban areas is the responsibility of the local authorities. But often lack of adequate financial resources impedes water supply efforts of the local authorities. The government, in its 1994-96 Development Plan intends to deal with this by restructuring the Local Government Loans Authority into a Municipal Development Bank to enhance the provision of financial resources to local authorities for development projects with priority given to water supply (9). The plan period is fast running out and yet little has been done in this regard.

Industries are mainly sited in urban and periurban areas increasing water demand in these areas and underscoring the need for reliable and adequate water supply. One way of encouraging industries to have their own water supply and thus relieve pressure on water supply in urban and periurban areas is to grant them tax exemption or rebates for the acquisition and installation of water supply, recycling and reuse technology.

Water use purposes. Under the Water Act, water use is required to be "reasonable as well as beneficial...". But the term "reasonable as well as beneficial" use is not defined. It can, however, be deduced from the different uses of water enumerated under s.35 of the Act in respect of which permits are required. Section 29 of the Water Bill (10) adds another water use purpose, namely: "disposal purpose, which expression means the disposal of industrial, agricultural, sewerage effluent and solid wastes from any other works or activities as may be determined by the Water Board". This provision makes water available for sanitary purposes, thus creating a safe environment for water supply, storage and use. Most periurban environment is epidemic-prone because of inadequate supply of water for sanitary purposes.

Water conservation imperatives. Conservation measures include the establishment of catchment areas under ss.14 and 22 and conservation areas under s.74 of the Water Act for the conservation of surface and ground water respectively. In such areas, special water quantity and quality measures can be taken in the interest of the public. But no such areas have been designated purposely for ensuring safe water environment in the periurban areas.

Unfortunately, the criteria for determining catchment area do not seem to directly include the support of aquatic environmental values. For example, swamps which provide natural purifying mechanisms for organic waste are being drained for agricultural purposes under ss.13 and 37 warranting the application of agro-chemicals which pollute water bodies. Kisumu town raw sewage which is being discharged directly into Lake Victoria due to poor maintenance of treatment plants, could have been channelled to Ahero and Yala swamps (which unfortunately have been drained without an environmental impact assessment process) for purification before discharge into the Lake. The Water Bill (s.11) fails to make significant improvement upon this as it only prohibits cultivation or cutting down of trees supposedly for agricultural purposes in designated river reserve and riverine forest reserve. This does not abate draining of such areas for purposes such as fish processing factories or tourist hotels.

Under the *Agriculture Act,* the Minister may prohibit, regulate or control land use activities such as livestock grazing or watering or the destruction of vegetation for purposes, *inter alia,* of maintaining "water in a water body within the meaning of Water Act" (11). Under ss. 4(1) and 6(1) of the *Forest Act* the responsible Minister may declare unalienated government land a "forest area" or "nature reserve" for the purpose of "preserving the natural amenities thereof and the flora and fauna therein." He may also excise such declared areas. These provisions preserve forest areas which serve as water catchment and purification mechanism. But the Minister's power of excision of "nature reserves" may negate the water retention and purifying function of the natural environment where it is arbitrarily exercised.

These pieces of legislation could be improved upon to achieve the desired goal of conserving and purifying water by incorporating in them the requirement of an environmental impact assessment for swamp drainage and excision of forest areas, and increasing the coverage of prohibited activities in riverine forest and river reserves under the Water Bill.

Determining safe yields of surface and ground water is important in the conservation of the resource. The *Seventh National Development Plan (SNDP)* provides for the assessment of the country's water resources (12). Under s. 21(2) of the Water Act, the authorities may require record keeping of the quantity and quality of water abstracted from any source by any person. Elaborate provisions for record keeping of flow, level, and quality of groundwater and test conducted by well drillers are made under ss. 51, 52, 53 and 73 of the Act. Surface water data can be obtained through the hydrographic survey provision under ss. 9(d), 20(b) and 100 of the Act.

Hydrographic survey and records of water quantity and quality by well constructors could signal where extra or more stringent conservation measures are needed. They may assist in determining safe yield of surface and ground water and thus provide a basis for informed water allocation decision. Efforts should, however, be made in Kenya to chart the aquifers and classify them according to their water yield, water quality and location, specifying the population they serve and the kind of use they are currently being put and should be put. Where such assessment is made, the Act or the regulation should embody the classifications and the use to which each class must be put. Such assessments are economically expensive but promote a rational utilization of water resources based on safe yield principle. This is a desperate need since in some basins, the "abstraction volume permitted by WAB appears to have already exceeded the available natural flow" (13).

The power of the WAB to prescribe measuring or controlling device for water abstraction under s. 28(1) of the Water Act could be exercised to reduce wastage. But it has been observed that some private water abstractors fail to install measuring device and compliance is hardly secured because of poor enforcement machinery (14). Section 24(a)(i) of the Water Bill introduces metering system. This will help reduce water wastage provided the government installs the meters because experience under paragraphs 44 and 45 of the Water (Undertakers) Rule which partly require individuals to install such meters shows poor compliance. In addition to metering system, rationing of water to minimize wastage is a proper water demand management principles which should be introduced to complement water supply management.

Disposal for Wastes, Sanitation, Water Pollution, and Water Quality Measures

Disposal of wastes and sanitation. UNEP Global Environment Monitoring Systems (GEMS) has reported water contamination by disease-bearing human wastes in many countries. This stems from poor system of disposal of human wastes including the discharge of untreated human sewage in bodies of surface and groundwater. It is reported that between 1980 and 1990, nearly a billion people had no access to safe drinking water (15). Clean water is needed not only for drinking but also for sanitation purposes.

In Kenya, cases of water pollution caused by waste disposal are on the rise. In Uasin Gishu district, for example, Eldoret town has more than 7,000 m^3 of effluent discharges within 10 km^2. The receiving water in River Sosiani has a low flow and dilution capacity and could not, particularly in the dry season, adequately handle the total BOD discharged by industries in the city. Furthermore, the two sewage treatment plants (one conventional and one lagoon plant) have not been functioning effectively as they receive amount of sewage beyond their capacities. Operational and maintenance problems have also contributed to poor sewage treatment and disposal. Consequently, sewage discharged into the receiving water of River Sosiani exceeds the allowable BOD concentration resulting in poor water quality (16). Given this scenario, the district stands a high risk of being affected by a wide spread epidemic of water-borne diseases. Kisumu area of Lake Victoria receives raw sewage from the municipal inefficient treatment plant that breaks down often. It has been reported that world-wide "unsafe water is implicated in many cases of diarrhoeal diseases, which as a group, kill more than 300 million people, mostly children, and cause about 900 million episodes of illness each year" (17).

Poor management of household, public and private toilets, commercial premises including market places and drainage systems could pose health hazards. The Public Health Drainage and Latrine Rules require every building, new or old to be drained. The local authority is empowered to enforce this requirement (paras 4-8). The authority may also require the drainage of a building that discharges wastes into a septic tank or cesspool to be connected to a municipal sewer where such is within 200 feet from the building (para.9). It is incumbent upon the owner of a premises to ensure efficient operation of the drainage (para 10). But the local authority may inspect drains, latrines, cesspools or septic tanks and, where appropriate, give order to the owner of the premises for their repair to abate or avoid a nuisance. Where the owner fails, he is guilty of an offence and the authority may undertake the repair and later recover the cost as a debt from the offender (para 11). Rain-water is prohibited from being piped directly into a closed drain (para.19) This is apparently to avoid excess water than the drain is designed for and thereby avoid flooding of latrines and other underground waste receptacles which could lead to pollution of water. For the same reasons, obstruction of drains is also prohibited: paras.13, 26-27.

Requirements for drain, water closet, slop-hoppers and urinal construction as well as waste-pipes and waste-water fittings, manner and procedure for excavation, gradients, construction materials and ventilation are provided in paras.28-47 and 50-61. Plans of drainage works are required to be deposited with a local authority and approved before the work commences (paras.64-76). This provision is to enable the authority to ensure that drainage works meet the required standard. The presence of substandard drains in the country, however, shows non-compliance and poor enforcement of this provision.

Septic tanks, storage tanks, sewage filter installation or any other works for the treatment, reception or disposal of sewage are required under para. 62 to be constructed of impervious materials so that they are watertight. Disposal of liquid or solid sewage including sewage effluent in a way that endangers the purity of water supply or causes nuisance is prohibited under para.63. It is also required that adequate and accessible latrines be provided for all users of existing public, domestic or business buildings including hotels. Provision of such facilities is a condition for building a new house (paras. 83, 84, 85 and 86). Latrines are to be covered and kept clean to avoid transmission of the wastes to food chain or water by the agent of wind or flies. Lack of adequate and accessible latrines lead to unsafe disposal of human faeces which enter water supply points and cause diseases like cholera or typhoid to those who drink such water. Latrines other than a water-closet or urinal, is prohibited from being constructed in connection with a building within 200 feet of a local government sewer on a suitable level and with sufficient water supply (para 91). Pail-closet latrines are not to be constructed within 10 feet of a kitchen or located within a dwelling house. The walls are required to be made of solid materials from bottom to top with adequate ventilation (para. 95). Under para. 96 construction of pit-closets by which human excreta is to be retained underground is required to be done with the written approval of the local authority who shall be satisfied that the soil condition and the intended location are suitable for such a purpose and who may prescribe further conditions for constructing same.

Sections 115, 116 and 117 of the Public Health Act prohibit nuisance and empower the local and health authorities to take lawful, and practical measures to prevent or stop nuisance in dwelling houses, trade premises or factories. They are to ensure the sanitation of such places. The Municipal Public Health Departments and the District Public Health Departments provide sanitary services, inspections and enforcements within the municipalities and the rural communities respectively. For example, in 1992, about 24,000 metric tons of refuse were deposited in the Eldoret Municipal Council's refuse tip. Nine big refuse containers and 300 bin containers were distributed to public and private premises for the collection of refuse for disposal (18).

Water plays a central role in disposing liquid wastes and in ensuring good sanitary conditions. Yet water for human consumption must be safe. The government reported operating and maintaining 683 water supply projects in 1992/93 fiscal year (19). Yet water supply is inadequate in urban and periurban areas due to increasing population and pollution. This has also affected sanitary conditions. The government however, plans to facilitate the development of sanitary facilities including the construction of reticulated latrines and septic water-borne facilities (20).

In its 1989-1993 plan period, the Uasin Gishu district, for example, planned *inter alia* to: 1) increase the number of households having access to safe water from 28% to 40%; 2) increase the number of household pit latrines and refuse pits from 60% to 100%; and 3) increase the number of health education outreaches from 157 to 800 a year. This goal was not achieved. The staff strength of the relevant implementing department, transportation for supervisory and inspection works, enforcement procedure and mechanisms, operation and maintenance of utilities were poor.

Water pollution control. Section 157(1) of the Public Health Act provides that where the Minister is satisfied that the irrigation or growing of crops in a township or within 3 miles of a township is unhealthy or insanitary, he may after consultation with the Agriculture Minister, prohibit such an activity and may cause the cancellation of the authorization given to carry out such activity. Urban and periurban farming particularly with fertilizers is known to contribute to pollution of water bodies which serve as water supply points leading to additional cost of treatment of such water before use. It seems clear that this provision has not been applied to check such polluting activities. Furthermore, the *Agriculture Act, Water Act* and *Water Bill* and other relevant statutes make no provision for a buffer zone for the purposes of application of pesticides, herbicides, fertilizers or other agricultural chemical in areas that may pollute surface water bodies. A buffer zone is a piece of land separating a surface water body from a farm where agricultural chemical is to be applied. It could stretch between 10 and 50 metres depending on the topography of the place. When the chemical is sprayed, rather than spill over into the water body, it will be absorbed in the buffer zone. It is important that such a provision be made in both the *Water Act* and the *Agriculture Act*.

Section 16 of the *Water Bill* gives the Minister the additional duty to "formulate guidelines and effluent discharge standards for the purposes of protecting water resources from pollution". The Minister can also determine, by order, the quality and location of effluent discharge into water bodies (21). This is consistent with government policy to ensure that both municipal and industrial waste water is treated to meet established water quality standards before being discharged into natural courses (22). While this is a move in the right direction, it however, presupposes that baseline studies of aquatic environment quality have been conducted to form the basis for determining informed guidelines and effluent discharge standards. It would also necessitate setting performance criteria for activities polluting the aquatic environment. This has not been done in most cases.

The hitherto inaction on the part of the government to formulate such guidelines and standards has compelled, for example, the Eldoret Municipality to make bylaws regulating the concentration level of substances that can be discharged into its sewers and subsequently into bodies of water (23). Unfortunately, the necessary baseline studies were not carried out and the concentration levels of the substances may have been determined arbitrarily. Again, monitoring may be difficult as there is hardly baseline information with which to compare pollution data.

Springboard provisions to implement such guidelines and standards already exist in the Water (General) Rules. These are, for example, provisions on the standard of the quality of effluent to be returned to a watercourse (24), prevention of contaminants from mining (25) and coffee industries from entering watercourses (26) and the prohibition of depositing any substance into watercourses deleterious to fish or fish habitat (27). Already, effluent discharge and standards for paper mills have been formulated and are enforced by the Pollution Control Unit of Ministry of Water Development (28). These however, are the general standards prescribed by the World Health Organization and there is hardly any effort to adapt it to local circumstances. Baseline studies which could have permitted this are hardly done.

Furthermore, a special discharge permit system for activities in the protected catchment and conservation areas should be adopted. Sections 10(c) and 16 of the Bill provide a good basis for this. Performance standards in relation to the desired water quality standard should be set and incorporated into the permit. This would presuppose that the background water quality in the areas would have been known so as to determine the differential degradation due to discharges. Based on this, the Allowable Discharge Levels (ADLs) of effluent should be set and specified in the permit (29). Section 35(b) of the Bill which provides for disposal of effluent and solid waste to be in accordance with the conditions in a permit could provide a springboard for incorporating ADLs into permits under the Water Act. It is however, important to set a more stringent water quality standard for groundwater than for surface water because contaminants dilution, biological breakdown of contaminants and volatilization of pollutants are slower in ground water (30).

There are other significant water protection provisions. Section 166(1) of the Water Act empowers the WAB to enter into any land for purposes of conserving and preserving water from pollution and for protecting the bed over which it flows. Section 68(1) prescribes some precautions for preventing groundwater contamination or pollution (31). The new s.158(1) of the Water Act created by s.58(a) of the Bill is significant as it provides a wider scope of prohibition of pollution by the use of the words "any body of water" (32). This contrasts sharply with the replaced provision which protects only water supply used for domestic, human consumption and food manufacturing processes. Water supply for drinking or domestic purposes is protected under s. 129(a)(b) and s. 130(1) of the *Public Health Act* through the prohibition of bathing, washing of clothing and the regulation of any such polluting activities (33) Subsection 130(2), however, provides that in regulating pollution of water used for drinking or domestic purposes, regard should be had to the interests of agriculture or industries. This *proviso* tends to weaken the water quality protection offered by the provisions and should therefore be expunged from the *Public Health Act.*

Section 27 of the *Bill* introduces remediation of contaminated or polluted water by providing that "the Water Board may declare certain water bodies as polluted and may prescribe measures for cleaning, rehabilitating or cleansing such water bodies". Unfortunately, clean-up standard is not prescribed. In principle, restoration should aim at attaining the standard of the natural environment prior to the contamination or pollution. This should be provided in the Bill. In practice, however, the ecological character of the environment, the nature of the pollutant or contaminant, and the best available and affordable technology are factors to be considered in determining whether or not such a standard can be attained in the remediation. This provision could arouse interest in the clean-up and restoration of already polluted aquatic environment. It is, therefore, a significant addition to environmental quality campaign.

Some scholars submit that "Kenyan anti-pollution legislation is adequate"(34). This is not totally correct. There is, for example, no direct legal provision for water protection for purposes of environmental values. Furthermore, there are no ambient aquatic environmental quality standards under the Water Act or other legislation of environmental relevance in the country. Special protection measures are limited in space, for example, in the "catchment areas" declared protected under s.150 of the Water Act or in the "conservation areas". And even in these areas there are no defined water quality standards (35). There are no performance standards for industries whose activities pollute water. Such standards are important as they ensure that pollutants discharged into a water body from a particular source do not exceed the acceptable limit. Criteria for determining the acceptable limit may include: 1) the natural dilution capacity of the particular water body, 2) its ecological characteristics, 3) the nature of the pollutants in question, and 4) other sources of pollution of that water body.

The provision of the *Water Bill* on guidelines and effluent discharge standards is positive but needs to be elaborated to specifically incorporate water quality standards as well as performance standards for polluting activities with respect to water quality. It would also require baseline studies of the present water quality to be conducted upon which such standards can be determined. Currently the Water Quality and Pollution Control Section of MOWD implements a nationwide water quality monitoring programme for both surface and ground water. It gathers baseline data on natural water quality as well as on the natural dilution capacity of surface water. Groundwater quality information is supposed to be supplied by borehole owners but only information on about 11% of the 10,000 boreholes in the country is supplied. This poor information gathering constrains a meaningful assessment of groundwater quality in the country. Regulation on securing such information from private well drillers will have to be tightened (36). The MOWD has to sink more groundwater observatory wells. But this would need financial and technical and personnel resources.

EIA and allied measures. Ensuring safe water environment through pollution avoidance or minimization measures cannot be realized without a legal provision for environmental impact assessment (EIA) and allied measures. EIA is a study of the effect of a proposed action on the environment; it compares various alternatives by which an objective may be achieved and identifies one that is the best from the point of view of the environment and economic; it is based on predictions of changes in environmental quality which may result from the proposed action, and serves as a decision-making tool (37). Environmental auditing is a systematic process of determining whether the management of a project is in compliance with regulatory requirements, national policy and set environmental management standards. It assures the public and the authorities that all measures have been taken to avert environmental hazards (38). Environmental monitoring is the process of determining where and when changes ought to be made in advance to foreclose a disaster.

There is no direct sectoral or umbrella legal provision for EIA, environmental auditing or monitoring. Accordingly, National Environment Secretariat's order for EIA was flouted by a government agency known as the Tana and Athi Rivers Development Authority (TARDA) (39). Unfortunately the Water Bill does not fill this ominous gap. The gap must be filled urgently as some studies have shown that Kenya will need many inter-basin water transfer schemes (40). The environmental problems in terms of negative effect on water quantity and quality and the aquatic ecosystem in general are enormous (41).

Housing and Drainage System Planning

Poor housing and drainage system planning could lead to health problems. In the municipalities where population density is high, requiring good house-spacings as well as equipping each house with acceptable sanitary facilities is important. Such facilities would include water system toilets, efficient wastewater disposal systems connected to municipal drainage or sewer systems and adequate ventilation. This will reduce the incidence of outbreak and spread of diseases. But only a formal housing programme could ensure this. While the government has made and continues to make efforts for the provision of adequate housing (42), the demand for decent housing outstrips the supply, leading to the establishment of informal, unplanned human settlement particularly in the urban and periurban areas. Furthermore, sizable portions of periurban areas are inhabited by low income earners who cannot afford decent, well planned houses with good sanitary facilities even where such are available. As a result, there are several unplanned houses with poor sanitation in such areas. The control of such developments is crucial for environmental health.

The Department of Physical Planning, in conjunction with District Development Officers, handle development planning. But they have thus far been unable to evolve long term plans that capture the principle of sustainable human settlement. According to one writer,

"A desirable district planning innovation would be to gather data for use by the ministries operating in the district in jointly developing a development strategy for the district. To date nothing of the type exists, for the District Development Plans are no more than catalogues or shopping lists put together by District Development Officers from information supplied by the various ministries. Very few districts have to date developed coherent district development strategies" (43).

Poverty

In developing countries numbers of the poor are increasing in both rural and urban cities. The linkage between poverty and health is pronounced. The urban poor live in congested slums and mosquito-infested areas. Garbage litters the settlement. Refuse and human waste disposal is not planned and therefore poorly handled. There is hardly clean water supply. Usually such areas are informal or illegal settlements and as such there is no plan to extend pipe borne water to them. Even where there is water supply, the poor often default in paying the charges. Yet water is needed for sanitation. Such services cannot be provided freely because of the enormous costs. Overall, sanitation in urban slums is menacingly poor posing tremendous health hazards.

Safe pit latrines such as ventilated improved (VIP) ones sited away from water tables, covered with slabs or other durable materials to keep away flies are not easily affordable by the poor. Unsafe pit latrines serving a large population are common. Hardly do the poor consult with the Public Health Department before siting such latrines because they fear being denied approval for not meeting the requirements for standard latrines. And often the substandard latrines collapse spilling human sewage on the road sides and into surface and ground water bodies. Poverty contributes to illiteracy which could impede the acquisition of health education.

Health Education and Culture

It is obvious that several health hazard scenarios develop because of ignorance on the part of the people. For example, knowledge of boiling water before drinking where the water is fetched from open ponds or streams or of disinfecting pit latrines and covering them with slabs could prevent a lot of diseases and sicknesses.

District health education campaigns have been slow due to certain constraints. The Uasin Gishu district, for example, had planned to increase home visits for purposes of health education from 396 a year as at 1989 to 9000 per year during the 1989-93 plan period. But this was not realized because of financial and personnel constraints.

As people realize that such education is for the prevention of otherwise deadly diseases, they will be willing to internalize the principles. Translating such basic health principles into practice will, overtime, crystallize into a norm in the society. The normative content of the law as opposed to "command and control" content should, therefore, be used as a tool for implementing sound health education. Generally, laws which do not reflect people's culture risk being ineffective. Such laws (for example, the Water Act and the Public Health Act) bring a radically different values which do not appeal to the people hence the use of "command and control" approach to secure compliance. On the other hand, laws which recognize people's values, use them as they are or refine them without perverting the culture persuade or wield compliance of the people. The normative contents of such laws are management oriented, appealing to the people's reasoning, way of life and overall well being. The people understand the message as being for their welfare and eschew any resistance. Therefore, reconciling cultural resistance with sound environmental health care should, from the legal angle and for the most part, be done normatively.

Recommendations and Conclusion

Demand pressure on public water supply projects can be reduced by requiring and encouraging industries to undertake their own water supply projects. Tax exemptions or rebates can be given to them for the acquisition or installation of water supply, recycling or reuse technologies. In this way adequate water can be conserved and made available to periurban settlers to promote, inter alia, good sanitary conditions. The Water Bill provision on the supply of water for waste disposal including sanitation purposes should quickly be made law and aggressively enforced particularly in the periurban areas. A sense of legal obligation to supply water for sanitation purposes could generate a serious commitment which otherwise will be elusive in this regard.

Criteria for designating conservation areas under the Water Act for special water quantity and quality measures should include the need to ensure adequate water quantity and safe water quality specifically in the periurban areas. Indiscriminate drainage of swamps and excision of forest areas which serve as natural water catchment and purifying mechanism should be checked by legally requiring EIA and allied measures for any such activity. There should also be a legal provision for the assessment and classification of surface and groundwater/aquifer sources to promote their rational utilization.

Application of agro-chemical in urban and periurban farms which lead to pollution of water bodies in these areas could be checked by introducing and imposing higher user fees and taxes in respect thereto. This should be complemented by providing for buffer zones in such farmlands in the Water Act and Agriculture Act. Substandard pit latrines which contribute to groundwater pollution can be checked by the government subsidizing the cost of constructing standard pit latrines where water closet latrines are not affordable. The law should provide for a legally protected consolidated revenue fund maintained by contributions from industries and government from which water supply, sanitary facilities and enforcement and compliance measures can be financed. This will help to avoid financial constraints which often constitute a set back in carrying out water supply and sanitary services.

The law should require baseline studies of major water bodies serving as water supply points. In addition, it should require the establishment of ambient as well as case-by case water quality standards. Performance criteria for each pollution source-activity, for example, by insisting, inter alia, on the installation of appropriate technology and by introducing and requiring pollution source-activities to have special discharge permits incorporating the limit of allowable effluent discharge should be provided in the law. This will make monitoring and enforcement easier.

The provision of the Bill generally prohibiting pollution in "any body of water" as opposed to the provision of the Water Act making such prohibition in respect to only water for human consumption and food processing is preferred and should be given a legal effect. The remediation provision of the Bill should also be given a legal effect but with an amendment to set clean-up standard to be the natural quality of the water (environment) prior to pollution. Section 130(2) of the Public Health Act which arguably derogates from the efficacy of water protection provision of s. 129(a)(b) and s. 130(1) of the Act should be abrogated or expunged.

There is a need for a review of planning law to capture the principle of sustainable human settlement which planners must be made to respect. However, financial and personnel resources needed to construct such human settlements are abysmally lacking particularly in the face of surging population and consequent increase in demand of decent houses. It is one thing for the law to make provisions for a sustainable human settlement and quite another to provide the resources needed to implement it.

The poverty and health education questions can be addressed appreciably by restructuring the social welfare system of the country. Financial and technical resources channelled directly to the poor, as well as good managers of such resources are needed if the poor are truly to be helped. Through education and social welfare campaigns which benefit the people directly in terms of, for example, providing affordable water supply and sanitary facilities, the people can be persuaded to discontinue cultural practices detrimental to safe environment. To help realize this goal, legal provisions should, for the most part, address the question of safe water environment normatively, that is, in a management-oriented way involving the understanding and participation of the people rather than largely using "command and control" or rule-oriented approach as is the case at present.

There should be a hybrid of the two approaches with the management oriented approach being a dominant component. "Command and control" approach will be complementary to deal with the few incorrigible people who would like to disregard the law.

References

1. G.C. Mutiso, "Management of Arid and Semi Arid Areas of Kenya" in A. Kiriro and C. Juma, *Gaining Ground: Institutional Innovations in Land Use Management in Kenya* (Nairobi: ACTS Press, 1989) p.76.
2. Republic of Kenya (Ministry of Water Development) *The Study on National Water Master Plan Sectoral Report "N" Environmental Conservation* (J.I.C.A) (Nairobi: The Ministry, 1992) p.N-1-7,8.
3. Republic of Kenya (Ministry of Water Development), *The Study on National Water Master Plan Sectoral Report* "M" Integrated Water Resources Development Planning (J.I.C.A.) (Nairobi: The Ministry, 1992) pp.172-177.
4. Eldoret Municipal Council, *Summary of Present Water Supply Position* (Eldoret: The Council, Oct., 1991) pp.1-2.
5. Republic of Kenya, *Seventh National Development Plan 1994-1996* (Nairobi: Government Printer) p.245.
6. *Water Act* Cap 372, ss.35 and 36. However, permit is not required for the abstraction or use of water for domestic purposes without the employment of works, or for the development of groundwater outside 100 yards from a surface water body. But either case is still subject to s.50 which provides a permit regime for groundwater withdrawal and s.74 which provides for a permit regime in areas declared "conservation area" by the Minister in respect of which special measures are applied.
7. *Ibid* s.98(1)(a)(b): operators aggrieved by the WAB measures except in cases of water shortage or declared emergency by the WAB, can appeal to the Water Appeal Board (now the Minister as per the Bill).
8. *Ibid* s.100.
9. Republic of Kenya, *Seventh National Development Plan 1994-1996* (Nairobi: Government Printer) p.246.
10. Bills No.12 published in Kenya Gazette Supplement No.51 of June, 1992 intended to amend the Water Act.
11. *Agriculture Act Cap 318*, s.48(1)(a)(i)(ii)(iii)(E).
12. Republic of Kenya, *Seventh National Development Plan 1994-1996* (Nairobi: Government Printer) p.200: government will intensify water resources assessment studies to improve their planning and exploitation.
13. Republic of Kenya (Ministry of Water Development), *The Study on National Water Master Plan Sectoral Report "P" Laws and Institutions* (J.I.C.A.) (Nairobi: The Ministry, 1992) at P-55.
14. *Ibid* at P-54.
15. World Bank, World Development Report 1992: *Development and Environment* (London: Oxford University Press, 1992) pp.45-47.
16. Ministry of Land Reclamation, Regional and Water Development, *Uasin Gishu District Annual Report 1992* (Eldoret: District Water Office, 1992) pp.2-6.
17. World Bank, World Development Report 1992. pp.48-49.
18. Eldoret Municipal Council, Public Health Department *Annual Report* for 1993, p.29.
19. Republic of Kenya, Seventh National Development Plan 1994-1996 (Nairobi: Government Printer) p.36.
20. Ibid p.246.
21. *Water Bill*, s.10(c) creating s.13(5) of the *Water Act*. And under s.45(b) of the *Bill* (creating new s.82(3) of the Act): such discharge is not permitted upstream or close to a water intake point unless the Director of Water Resources advises otherwise.
22. Republic of Kenya, *Seventh National Development Plan 1994-1996* (Nairobi: Government Printer) p.200.
23. See for example, The Municipal Council of Eldoret (Trade Effluent) By-Laws, 1991 (Legal Notice No.541 issued under the Local Government Act Cap.265), ss.3, 7(2)(3) and the schedule thereto.

24. *Water (General) Rules*, para.72(1).
25. *Ibid.* 73(1).
26. *Ibid.* para.74.
27. *Ibid.* paras.77, 80.
28. *Laws and Institutions Sectoral Report "P"* ,p.P-65.
29. This is already the case in Michigan as per: *The Michigan Water Resources Commission Act (MWRCA) (Public Act 245 of 1929 as amended 1980)* Part 22 Rules, s.323.7. Arizona State has similar measures as per: Arizona Department of Environmental Quality, *State of Arizona Groundwater Protection Strategy* (Arizona: The Department, 1989) at 29.
30. Michigan Department of Natural Resources, *State of Michigan Groundwater Protection Strategy and Implementation Plan* (Michigan: The Department, 1989) at 22.
31. Other pollution measures include confining salt water in the strata where it was found to avoid its escape to other water bearing formation: Water Act, s.58; and prohibition of rubbish dump in defective wells without prior authorization, s.59.
32. *Ibid* s.58(a).
33. Cap 242, revised 1986.
34. D.A. Caponera et al., *Water Law in Selected African Countries* (Rome: Food and Agricultural Organization of the United nations, 1979) at 136.
35. *See for example*, the subsidiary legislation LN311/958; LN374/1964 in which certain areas were declared "conservation areas".
36. *Laws and Institutions Sectoral Report "P"* ,p.P-63-64.
37. Ahmad, Y.J. and Sammy, G.K., *Guidelines to Environmental Impact Assessment in Developing Countries* (London, Sydney: Hodder & Stoughton, 1985) at 1-2.
38. Thanh, N.C. and Tam, D.M., "Environmental Protection and Development: How to Achieve a Balance" in A.K. Biswas and S.B.C. Agarwala, *Environmental Impact Assessment for Developing Countries* (Oxford, London: Butterworth & Heinemann Ltd., 1992) at 13.
39. K.K. Orie, "Constitutional Arrangements and Sustainable Development" in C. Juma and J.B. Ojwang ed. *In Land We Trust: Natural Resources and Kenyan Constitution* (Nairobi: ACTS Press, 1995).
40. *Law and Institutions Sectoral Report "P"* ,p.P-49.
41. Republic of Kenya, (Ministry of Water Development), *The Study on the National Water Master Plan, Main Report Vol.I: Water Resources Development and Use Plan Towards 2010* (Nairobi: The Ministry, 1992) at 132-134.
42. Republic of Kenya, *Seventh National Development Plan 1994-1996* (Nairobi: Government Printer) p.247.
43. Mutiso, supra, p.83.

COMMENTARY ON *THEME V.*

Re Paper 12

Ms. A. Almedon (LSHTM) remarked that considerable prestige is associated with being a benefactor or prime mover in securing a village water supply in rural Africa. Wells are often named after these people. This endorses Mr Hamlin's point that water supplies are not built for health reasons alone.

Ms. J. Kunguru commented that much of the underlying reasoning for constructing water supplies mentioned by Mr Hamlin rang true of many East African communities. As such they are, in addition to our perceived scientific values helpful considerations for community workers to use in development projects.

Dr. R. Swiderski also commented upon the usefulness of these alternative considerations. He raised the point that there are many similarities between cultures of the healing properties of water, which also could be drawn from a cosmological view point. In separating water from these belief structures we do not treat water as a component element but more as a separate entity.

Mr J.H. van Doorne (Moi University), emphasised the point that approaching a community could only be made honestly, attempts to introduce a hidden agenda or a premeditated solution would not be successful.

Mr R. Watts (Moi University) questioned wether western authorities had gone "overboard" on the issue of water quality, asking why there is this constant pressure to raise standards, particularly when we consider how very little of the water in a piped supply is actually drunk.

Mr C. Hamlin responded by saying that this may be more an issue of the motives for privatisation of supply and the need to show returns to shareholders. *Mr S. Cairncross* (LSHTM) added to this response by questioning what information was in the best interests of the consumers. On the one hand it may be difficult to get accurate water quality data into the public domain, whilst on the other its effect may be excessively disruptive. For example should the volumes and prices of water sold through vendors or water kiosks be known what would be the effect in say Nairobi.

Mr M. Woodhouse also mentioned that defensive or strategic location represents a further possible motivation for securing a water supply.

Re Paper 13

Mr. R. Swiderski commented that cultural differences must also be considered in ascribing different reactions to water and sanitation projects. He pointed out that Jakarta is a predominantly Islamic city, Sao Paulo predominantly Catholic. However it was recognised that there are fluxes in culture.

Dr C. Walekhwa said that in Kenya the provision of public toilets is generally considered to be a failure, through misuse and vandalism, he sought other participants experiences on the matter. *Mr G. McGranahan* responded by saying that in Accra some public toilets were successful where a fee was charged to cover the attendants salary and maintenance. But there are difficulties in determining who should pay that fee. For example should children pay?

Ms J. Kunguru added that there are successful public toilets working in Kampala, Uganda. For a small fee toilet paper is provided and a salary paid to the attendant, whilst *Mr Haraka* mentioned seeing a similar system working in Scotland, there was however a vandalism problem.

Some open discussion was fuelled by these comments upon who should be charged a fee and for what. Some people who do not have a latrine in their house may want to simply empty the contents of their household receptacle at the public toilet. It was also mentioned that there are examples of the fees collected at public toilets not being used for their upkeep. Public toilets for residential areas were seen as a last resort where the provision of household level toilets is not possible. There are also many cultural taboos to using communal public toilets.

Dr C. Walekhwa raised the issue of how should we consider space as an indicator of environmental health in residential areas, unfortunately no time was available for discussion.

Re Paper 14

Ms. E. Hurtado (Institute of Nutrition of Central America and Panama) mentioned that Participatory rural appraisal (PRA) and rapid rural assessment (RRA) are often though to be synonymous. The distinction is very important as they are techniques used to gain very different types of information.

Mr M. Woodhouse mentioned that participatory approaches have been very popular, yet recently there is more talk of demand driven approaches and asked if this did not run the risk of contradiction at the community level.

Mr J.H. van Doorne (Moi University) felt that even though we try to work with an open attitude we also harbour undisclosed assumptions. He also raised the issue that institutionalisation of a participatory approach was important at the senior government level, yet much of the effort appears to be going on at the grass roots level. *Mr. J. Thompson* responded by mentioning work done in the Ministry of Agriculture in Kenya where participatory approaches were evaluated and incorporated at the senior level. In response to the existence of undisclosed assumptions Mr Thompson referred the questioner to writings on " New professionalism".

Astier Almedom (LSHTM) introduced a handbook on hygiene evaluation procedures, and said that number of her colleagues in the Department of Epidemiology and Population Sciences at the London School of Hygiene have long been concerned with the evaluation of water supply and sanitation interventions. For instance, Sandy Cairncross and others provided support to Gunnar Schultzberg in 1981-82 when he prepared the Minimum Evaluation Procedures, MEP, Handbook (WHO, 1983) to assess operation, maintenance and use of facilities. Ten years later Ms. Almedom started work on a similar Handbook, the Hygiene Evaluation Procedures, HEP, focussing on the assessment of hygiene practices in the context of water and sanitation projects.

This Handbook is primarily aimed at field personnel in water supply, sanitation and health/hygiene education projects. Its approach has been based on the principles of problem based learning, in particular, Kolb's Experiental Learning Theory (Kolb, 1984). The HEP Handbook aims to assist its users to design and conduct systematic assessments of hygiene practices using a range of methods/tools which can facilitate relativelyquick and thorough investigations. The methods and participatory tools (such as those mentioned in John's paper) originate from anthropology and other related disciplines, and they have been field tested with the collaboration of a number of projects, including one in Kenya.

The language of the HEP Handbook reflects that of practitioners rather than academic researchers. For instance, the term "problem identification" is used instead of "defining research questions"; "information gathering" instead of "data collection"; "review of information" instead of "data analysis"; and so on. Data analysis is explicitly linked to investigative activities. A quick look at the Table of contents of the Handbook shows that it is a practically oriented document. It is worth pointing out that the training of project staff is a key component of the Handbook. Both initial/introductory training, with a mix of class-room based sessions and field visits, and on-the-job training during the actual conduct of the study is provided. This provides the potential users of the Handbook with the opportunity to learn through experience in a relatively short period of time. The time required for a project to conduct its own systematic assessment of hygiene practices among the served communities varies according to the project's capacity (human and material resources). So far, we have found that given the necessary resources, one assessment (a cycle of hygiene evaluation) can be conducted in a matter of weeks, 6-8 weeks, including a two week period of pre-planning and training.

The HEP Handbook was developed through a series of field trials of the methods and tools in Kenya, Tanzania and Ethiopia. Field testing of this draft is about to start in India and Afghanistan.

Re Paper 15

Mr M. Karama (KEMRI) made the point that since it is very difficult to adopt WHO standards for water quality in the local context how should realistic standards be set. *Mr. G. Schultzberg* mentioned that the WHO sets guidelines for standards, not standards themselves. National authorities can use this guidance to establish their own standards. Tanzania for example has a number of incremental standards, which aim over the long term to reach the WHO guidelines. *Mr. Orie* responded by stating that the law has to be related to scientific fact and the law will fail if it does not fit the local context. Accurate data are required before we can legislate. There are many constraints to obtaining adequate data on the local levels of contaminants in water.

Mr J. Kwonyike (Moi University) acknowledged the authors response but also raised the issue of pragmatism. Profiteering industrialists want to see a grater return on a lower investment. The political will has to be there to apply the law and in this case precludes the need for scientific data. The law is impeached if there is no will to enforce it.

Mr Haraka (Nakuru Municipal Council) mentioned that industries in Nakuru are obliged by law to present an environmental Impact assessment before any new ventures can be approved. This situation being precipitated by the worsening condition of Lake Nakuru an area of great environmental importance.

SUMMING-UP OF CONFERENCE DISCUSSIONS

Sandy Cairncross and Peter Nyarango

Dr Cairncross summarised the conference discussions. Underlying much of the debate, there had been a range of views regarding the relative importance of water quality on the one hand, and of hygiene and water use on the other, as determinants of health. There was a consensus, however, that an understanding of human behaviour was essential to any attempt to improve the situation. The provision of good quality water does not ensure that the water will be used in the most beneficial way, or even be used at all.

We need therefore to understand local perceptions and knowledge, and be aware of the fact that water is an important cultural vehicle. Even where water has largely been transformed into a commodity, as in Western industrialised societies, bottled water has become a prestige good, showing how this substance can never be robbed of all cultural significance.

There was a contrast of emphasis in the discussions on the peri- urban environment, between the academic researchers, many of them foreigners, and the practitioners working in public health in city administrations. Whereas the academics focussed largely on water- related issues, the practitioners were more interested in sanitation, drainage and land tenure.

There was general agreement on the need for dialogue between those who plan and implement projects and the communities which those projects aim to serve. Remarkably cohesive institutions often exist at the level of the neighbourhood of 3,000 to 5,000 people in Third World cities. However, a number of references were made in the discussion to the conflicts of interest which may exist among the inhabitants of a neighbourhood; between landlord and tenant, between upstream and downstream neighbours, and so on. The dialogue needed is not with just one community or leader, but with a number of different elements.

The presentations also underscored the importance of endemic disease, of long-term trends in pollution and of underlying causes, and not only the unusual or sudden events such as epidemics and their immediate or precipitating causes.

Another recurring theme was the so-called "KAP gap"; the fact that new knowledge does not necessarily lead to the corresponding changes in attitudes or practice. It became clear that the gap between knowledge and practice has complex reasons, and is found not only among the uneducated and the poor, but also among the professionals who manage urban environmental improvement programmes. Those who acknowledge the need to consult the community, for example, do not always do so.

There were a number of appeals from the practitioners for new technical solutions; however, many of the technical options have already been well researched, though the resulting knowledge has not been effectively disseminated to those who could use it.

The goal of making environmental improvements sustainable remains an elusive one, and more research is needed to discover the determinants of sustainability. For example, the currently fashionable view that demand-led projects are more sustainable needs to be tested in practice. "Development archaeology", in which projects are examined several years after their conclusion, seems to be a promising way to study such questions. We have much to learn from history.

The data presented at this conference on the movement of bacteriophages in groundwater are of great concern, as the rates and distances involved far exceeded our expectations. Are these results typical of other parts of the developing world, and what are their health implications?

There is room for more research on almost all the issues that have been touched in this conference, but another important conclusion for me is that there is also a great deal of room for exchange of information between cities, within Kenya, with neighbouring countries, and with other parts of the world. The conference has been for many of its participants a great opportunity to make contacts and share experiences. It is to be hoped that these contacts will spur new research initiatives by the academics, and help the practitioners to implement better solutions in the field.

OPEN QUESTION SESSION

Mr.C. Hamlin (University of Notre Dame) felt that so far little has been said on the responsibilities of governments in this arena. It would be foolish to consider achievements when there is not full governmental participation.

Mr. J.H.van Doorne (Moi University) On the point of defining responsibility said it should be recognised that there is great mobility in periurban populations. You may be dealing with different people from one month to the next. People in periurban areas will tend to move on when they perceive that an environmental problem is beyond their control. They have the option to leave and live somewhere else.

Ms. A. Almedon.(London School of Hygiene and Tropical Medicine) recognised that there is also a gap between our research findings and subsequent implementation of activities. The KAP gap does not exist solely in the community.

Mr. G. Schultzberg commented that the institutional bureaucracies we work with have been built up over the last 20 years or more. Our present shift of focus to NGO led development is a much more recent phenomenon. It will take time to evolve and we have to consider the implications then of creating a gap between institutions and communities.

Mr. G. Araka.(Nakuru Municipal Council) noted the emphasis placed on microbiological contamination of water in this conference and stressed we should not ignore chemical and heavy metal pollution.

Dr. C. Walekwa (Eldoret Municipal Council) said that drainage was perhaps being perceived as a lesser evil, the provision of water is more heavily stressed. We have to consider drainage of waste water and its benefits more thoroughly.

CLOSING

Proceedings will be prepared and distributed to the participants. Prof Nyarango forwarded in his closing speech a vote of thanks to SIDA, Linköping University and the London School of Hygiene and Tropical Medicine.

POSTER SESSION

WATER QUALITY HYGIENE BEHAVIOUR AND CHILDHOOD DIARRHOEA - IS THERE A RELATIONSHIP?

Asheesh Aggarwal, James Gikandu Kamau, Patrick F. O. Sawa, Lulu Nyawira Oguda
Medical students, Faculty of Health Sciences, Moi University, Eldoret

Background. Faecal contamination of wells may be attributed to poor hygiene behaviour and is presumably a major contributor to childhood diarrhoea in periurban setting such as Langas in Eldoret.

Aim. In a longitudinal prospective study we investigated the influence of the mothers hygiene behaviour and the feacal coliform counts of well water and their detriments on childhood diarrhoea

Introduction. In Langas and other peri-urban areas in many developing countries, prevalent conditions such as over crowding and poverty ease the transmission of diarhoeal diseases. Feacal contamination of drinking water is a major route of spread of these diseases which are a major cause of childhood mortality in poor countries. Five million children die of diarrhoeal diseases every year.

Methodology.
- Observations
- Interviews
- Focused group discussions
- Water sampling and bacteriological analysis

Results:
- There was significant correlation between hygiene index and bacteriological quality of water.
- Incidence of diarrhoea was significantly higher in the less hygienic compounds.
- The distance between well and latrine did not influence the bacteriological quality of water.
- No correlation between the water quality and the incidence of diarrhoea.
- Latarines, surface intrusion and water drawing process were shown to be significant sources of well water contamination.
- Knowledge on the appropriate hygiene behaviour was almost universal, the actual practices were minimal.

Conclusion:
- Hygiene behaviour influences the incidence of childhood diarrhoe though not only through water.
- Thus improving hygiene behaviour using preventive health education strategies on the existing socio-cultural values may reduce childhood diarrhoea.
- There is need for studying socio-cultural implication and economic costs of improved behaviour.

Suggestions:

1. Health education as a means of health promotion should be tailor-made to needs of local community.

2. The residents should be made aware of the health risks that exist without proper handwashing, unhygienic use of latrines and importance of sanitation and effective water storage and use.

3. The Eldoret municipal council is advised to plan some long term solutions on water supply and sanitation services in Langas and other peri-urban areas.

4. The Faculty of Health Sciences should continue promoting similar studies to reach definite and scientific decisions on water quality, hygiene behaviour, diarrhoea and its relationship so as to achieve health for all by year 2000.

Discussion.

1. Hygiene behaviour is an important influencing factor of childhood diarrhoea in Langas.
2. Mean incidence of diarrhoea per child per 18 weeks was 1.6 - therefore 4.6 episodes per year, but Snyder and Merson (1987) found 2.2 episodes per year suggesting diarrhoea is a principal source of morbidity in Langas.
3. We found no correlation between well water quality and distance of wells from latrines because of other sources of contamination such as handwashing, poor disposal of faeces and household refuse and geomorphology of the area.
4. The mean feacal coliform count for 8 wells was 300/100ml. In this perspective, the WHO Guidelines are farfetched therefore the need for a locally acceptable figure.

Food For Thought

During our study, a lady posed this question.
" *If I am not sure my children will have a meal tomorrow, what's my business boiling water?*"

CONFLICTING INTEREST BETWEEN LANDLORDS OVER STORM- AND WASTEWATER DISPOSAL

Anthony Macharia, Safe Water Environment programme and Moi University

Introduction. The study was carried out in periurban Eldoret, Huruma and Langas. Few people have plot connections for wastewater from washing and bathing. The alternative is disposal on ground or in stormwater drainage.Most vegetation was destroyed during settlement. This in combination with rocky or sloping areas in Huruma has contributed to poor drainage.The partially installed drainage system by the 3rd Kenya urban development plan did not solve the disposal problem. The inadequate existing system has been exposed to abuse by local community. The ditches are either collapsed or partially blocked by refuse, silt, stones and plastic materials.

Aim. To investigate factors influencing the present status of storm- and wastewater disposal, from observations and interviews of informants in the field and municipal staff.

Factors influencing the situation:
- lack of discipline over house building and clearing of vegetation
- lack of community participation
- lack of cooperation from the municipality department has affected the hygienic situation
- lack of proper planning, community participation and cooperative efforts has affected the relationship between landlords
- present local community efforts in solving drainage problems.

Results and discussion:
- Conflicts regarding drainage problems are seasonal. Conflicts are prominent during the rainy season.
- Arbritation between landlords involves affected parties together with the chief organization, including the headmen and selected elders.
- Conflicts on wastewater disposal between landlords have occurred due to measures by the 3rd Kenya urban development project that resulted in legal suits
- Pending legal suits in law courts and lack of planning hinders arbritation between affected landlords
- Most houses are permanent. More space for roads and drains are difficult to create.

Conclusion:
- Provision of adequate surface water facilities is essential to the health of any community
- Adequate surface water facilities is a responsibility of municipal departments and the local community
- Negotiation between landlords regarding drainage problems are solved by mutual understanding and cooperation

WELL WATER AND TAP WATER USE IN PERIURBAN AREAS OF ELDORET

Jerry Okal
Safe Water Environments programme and Moi University, Eldoret

Introduction. The above study was carried out in periurban areas of Huruma and Langas. The study involved aspects of tap water and well water use and the social and economic aspects.

Objective/Aim: To get knowledge on peoples:
- Preferences of well water and of river water.
- Understanding the use of wells, springs and rivers in periurban areas vis-a-vis rural areas.
- Judgements and perceptions concerning the difference between safe and clean water.
- To know the amount of water drawn from both wells and taps.

Methodology:
- Observations
- Interviews
- Literature review

Results:
- Water is judged by appearance, taste and smell.
- River water considered clean because it is flowing.
- There is only slight differences on the use of wells, rivers and springs here in town and in the rural areas.
- Most people could not differentiate between clean and safe water.
- Tap water is believed to be naturally clean. Most people are unaware where this water comes from.
- Those with tap water draw this water early in the morning and late in the evening.
- In the dry season water is drawn from wells which never dry up.

Discussion:
- Protecting well water is more difficult than protecting air or surface water.
- Public concern for groundwater is greater than other environmental issues.
- The municipal authority should ensure safety of the wells.
- Public ownership of water supplies is important, this is not a question of socialism.
- Piped water would be improved by rehabilitating consumer connections.
- Improving water billing to reflect the actual consumption.
- Chlorine and alum should be made available to plot owners.

Conclusion:
- Well water is extensively contaminated, this is as a result of human related activities and environmental factors.
- Publicity campaigns to reduce water consumption and contamination have been successful in a number of countries. It is likely that if socio-cultural preferences are taken into account similar campaigns would prove effective in developing countries.
- Information on water supply system could be coupled with health education in schools, "barazas" etc.
- Sound groundwater strategies are necessary in areas where ground water source is still exploited.

CAN THE LANDLORDS CHLORINATE THEIR WELLS?

Otieno Auko,
Veterinary Research Laboratory, Eldoret

Objective. To study chlorination of well water by landlords to evaluate it's efficiency.

Methodology. Firstly, faecal coliforms in the well water were counted by membrane filtration. Hypo-chlorite was added to the well water, as done by the landlords. Total chlorine was determined by using DPD4 tablets and a Lovibond chloriometer. Sampling of well water was done after 30 min and continued for 10 days.

Findings: Well water quality (Table below) showed heavy contamination by faecal coliforms. Faecal coliforms disappeareded totally after chlorination, but reappeared after 3 days when the chlorine load was low or non-existent.

DAY	WELL 1 FC per 100 ml	WELL 1 Chlorine in mg/litre	Well 2 FC per 100 ml	Well 2 Chlorine in mg/litre
-1	220	>5	500	3.5
0	0	2.5	0	2.5
1	0	–	–	1.3
2	–	–	–	–
3	–	–	–	–
4	120	0	250	0

CONCLUSIONS

- Chlorination is the most effective method of water disinfection so long as proper concentration of hypochlorite compound is used and continuously applied. This should be done together with alum treatment to encourage the user.
- The shock chlorination is effective on faecal coliforms but gives unhealthy water and landlords should be guided.
- There should be periodical sampling of the wells to test for indicator bacteria as a quality control of the water.

HAZARDOUS BEHAVIOUR THAT MAY LEAD TO WATER CONTAMINATION

Bernard M Mundia
Safe Water Environment-programme and Moi University

Introduction. In a local brewing and merchandising plot endeavours were made to identify some behaviours that could contribute to water contamination. Local brew may influence behaviours that in turn contribute to water contamination and diarrhoea among children.

Aim:
- Identify dangerous behaviours that could lead to water contamination
- Assess the extent to which local brew influences water contamination

Methodology:
- Observations
- Laboratory test of water samples drawn from the well.

Results of sampled water.(4 occasions)

FC/100 ml	TC/100 ml	FS/100 ml
32	6000	2800
3	26	62
1200	1300	300
2300	5000	2500

Hazardous behaviour
- Poor protection of wells; open - no lid
- Washing near the well
- Young children draw water and play - not taken care off; also drink local brew
- Dropping of objects into wells
- Animals around water source
- Neither boiling of deinking water nor chlorination of wells
- Dipping of fingers into water containers

Poor defecation practices:
- Inability to dispose faecal material in latrine
- Open field defecation
- Faecal material around water source
- No hand-washing after defecation.

Conclusions and Statement:
- A number of hazardous human behaviours may contribute to water contamination
- Local brew occasionally influence human behaviour which in turn influence water quality
- An understanding of peoples practices and behaviour is essential in order to find possible solutions

TENANT - LANDLORD NEGOTIATIONS: IS "VOTING BY THE FEET" AN IMPEDIMENT TO CHANGE?

Maurice Kongongo
Safe Water Environments Programme and Moi University

Introduction and aim. This study was conducted in Langas; a periurban area in Eldoret with about 40 000 inhabitants. This is an area for low-income earners.
The objectives were to elucidate peoples perceptions, attitudes and knowledge about hygiene and hygienic behaviours. Emphasis was laid on cooperative efforts among tenants and landlords.

Methodology:
- participant observation using structured and unstructured observation protocols.
- interviewing using questionnaires (structured) and unstructured questions
- random selection of 14 respondents (10 tenants and 4 landlords)

Results:
1. No written agreement when a new tenant occupies a house. Only receipts are issued to certify payment of rent. Assumptions were that all tenants know what should be expected of them.
2. No legal actions in most cases are taken against rent defaulters. Household items might at times be taken instead.
3. Tenants may "vote by the feet" if not satisfied with environmental conditions. They may be willing to pay more for cleaner surroundings
4. Provision of sanitary facilities is regarded as the responsibility of the landlord. Tenants kept the closest area around their house clean.
5. It was found that the sanitation situation depends more on the good will of the landlord than his actual presence in the compound.
6. No negotiation occurred about latrine emptying. "The landlord has a duty to do it."

Conclusions:
- The landlords operate without any legal guidelines. Conflict resolution between them becomes difficult.
- Many tenants just walk away when they are not satisfied with the water and sanitation situation.
- This may impede improvements because landlords´ sources of finance have been cut.

List of Participants

Aggarwal, Asheesh
Moi University -Med Students
P.O. Box 4606
Eldoret

Akonga, Joshua
Moi University
P.O. Box 3900
Eldoret.

Allestam, Gorel
Swedish Institute of Infectious Disease Control
S-105 21 Stockholm
Sweden.

Almedom, Astier
EPS, London School of Hygiene
Keppel Street
London WC1E 7HT
England.

Araka, O. Glenn
Municipal Council,
P.O. Box 124
Nakuru.

Auko, Otieno
Veterinary Research Lab.
P.O. Box 450
Eldoret.

Ayuku, David
Moi University - Fhs
P.O. Box 4606
Eldoret.

Balfour, Nancy
AMREF
Environmental Health Progr.
P.O. Box 30125
Nairobi

Burgess, J. Stephen
CPK Diocese Eldoret
P.O. Box 3404
Eldoret

Buziba, Nathan
Moi University - FHS
P.O. Box 4606
Eldoret.

Bradley, David
London School of Hygiene
Keppel Street,
London Wc1e 7ht
U.K.

Caincross, Sandy
London School of Hygiene,
Keppel Street,
London WC1E 7HT
U.K.

Chege, Ephantus
KMTC
P.O. Box 30195
Nairobi.

Chesang, K. Wilson
Municipal Council,
P.O. Box 40
Eldoret.

Chepkuto, P.K
Moi University
P.O. Box 3900
Eldoret.

Drangert, Jan-Olof.
Linkoping University
581 83 Linkoping
Sweden.

Doorne, J. H. Van
Moi University,
P.O. Box 3900,
Eldoret.

Edebe, W.M
Moi University
P.O. Box 3900
Eldoret

Hamlin, Christopher
University Of Notre Dame
Notre Dame, In 46556
USA

Hurtado, Elena
Institute of Nutrition of Central America and Panama
19 Av. 8-44 Cona 15
Vista Hermosa 1
Guatemala

Kamar, K.K
Moi University - Fhs
P.O. Box 4606
Eldoret

Kamau, M.S
Moi University
P.O. Box 3900
Eldoret

Kamau, G. James
Moi University (Med.Stud)
P.O. Box 4606
Eldoret

Karama, M.
KEMRI
P.O. Box 20752
Nairobi

Kareithi, J.N.
Moi University
P.O. Box 3900
Eldoret

Kariuki, A. Macharia
P.O. Box 13231
Nakuru

Kemboi, J. Naomi
CPK Diocese Eldoret
P.O. Box 3404
Eldoret

Kibochi, Njau
KEMRI
P.O. Box 19464
Nairobi

Kong'ong'o, Maurice
C/O Erastus Obiayo
P.O. Box 6573
Kisumu

Kunguru, Julia
World Bank, RWSG,
P.O. Box 30577,
Nairobi

Kwonyike, Joshua
Moi University
P.O. Box 3900
Eldoret

Mackambo, John
Moi University
P.O. Box 3900
Eldoret

Maithya, Harrison
Moi University
P.O. Box 3900
Eldoret

Manyim, Geofrey
CPK Diocese Eldoret
P.O. Box 997
Eldoret

Makayoto, Moses
ICIPE
P.O. Box 30772
Nairobi

Mengech, arap
Moi univesity - FHS
P.O. Box 4606
Eldoret

Mgana, Shaaban
Ardhi Institute,
P.O. Box 35176,
Dar-es-Salaam,
Tanzania

Mohammed, Alia.
Municipal Council,
P.O. Box 371
Malindi

Mundia, Bernard
P.O. Box 1287
Maragoli

McGranahan, Gordon
Stockholm Environ. Institute
P.O. Box 2142
S-103 14 Stockholm
Sweden

Munyirwa, Kinya
Osienala (Friends of L. Victoria)
P.O. Box 4580
Kisumu

Mutua, Eda
Moi University
P.O. Box 3900
Eldoret

Nangalama, R.
Ministry Of Health
P.O. Box 9085
Eldoret

Ndege, K. Samson
Moi University - FHS
P.O. Box 4606
Eldoret

Ngare, Duncan
Moi University - FHS
P.O. Box 4606
Eldoret

Nyarango, Peter.M.
Moi University - Fhs
P.O. Box 4606
Eldoret

Olendi, Robert
Moi University,
P.O. Box 3900
Eldoret

Olubero, Timothy
Kakamega District
P.O. Box 750
Kakamega

Oluoch, O. Francis
Bungoma District Health Office
P.O. Box 383
Bungoma

Okal, O. Jerry
P.O. Box 206,
Homa Bay,
Kenya

Oundo, Joseph
KEMRI
P.O. Box 19464
Nairobi

Opata, G.
Moi University
P.O. Box 3900
Eldoret

Orie, Kenneth
Moi University
P.O. Box 3900
Eldoret

Otieno, A. Dismas
Moi University,
P.O. Box 3900,
Eldoret

Rotich, Reuben
Eldoret Municipal Council
P.O. Box 40
Eldoret

Sang, K. Willie
KEMRI
P.O. Box 19464
Nairobi

Sawa, Patrick O.
Moi Univesity (Med.Stud)
P.O. Box 4606
Eldoret

Schultzberg, Gunnar
P.O. Box 30600
Nairobi

Suleiman, Saidi M
KEMRI
P.O. Box 19464
Nairobi

Sundgren, Margaretha
SIDA,
S-105 25 Stockholm,
Sweden
Stenstrom, Tor Axel
Swedish Institute for Infectious
Disease Control
S-105 21 Stockholm
Sweden

Swiderski, Richard
956 Old Quarry Rd South
Larkspur Ca 94939
U.S.A.

Thompson, John
International Institute for
Environment and Development
3 Endsleigh Street,
London WC1H ODD
United Kingdom

Walekhwa, Charles
Municipal Council,
P.O. Box 40
Eldoret

Wamola, Isaac
Moi University - FHS
P.O. Box 4606
Eldoret

Watts, Ronald
Moi University
P.O. Box 4606
Eldoret

Watts, Teresa
Moi University - FHS
P.O.. Box 4606
Eldoret

Wabwire, Anthony
Moi University
P.O. Box 3900
Eldoret

Waweru, Nelson N.
Ministry of Health
P.O. Box 30016
Nairobi

Woodhouse, Melvin
P.O. Box 9200,
Dar-es-Salaam,
Tanzania